Fundamentals of Integrated Coastal Management

The GeoJournal Library

Volume 49

Managing Editors: Herman van der Wusten, University of Amsterdam,
The Netherlands
Olga Gritsai, Russian Academy of Sciences, Moscow,
Russia

Former Series Editor:
Wolf Tietze, Helmstedt, Germany

The titles published in this series are listed at the end of this volume.

Fundamentals of Integrated Coastal Management

by

by

ADALBERTO VALLEGA
International Centre for Coastal and Ocean Policy Studies, ICCOPS,
University of Genoa,
Italy

KLUWER ACADEMIC PUBLISHERS
DORDRECHT / BOSTON / LONDON

A C.I.P. Catalogue record for this book is available from the Library of Congress

ISBN 0-7923-5875-9

Published by Kluwer Academic Publishers,
P.O. Box 17, 3300 AA Dordrecht, The Netherlands.

Sold and distributed in North, Central and South America
by Kluwer Academic Publishers,
101 Philip Drive, Norwell, MA 02061, U.S.A.

In all other countries, sold and distributed
by Kluwer Academic Publishers,
P.O. Box 322, 3300 AH Dordrecht, The Netherlands.

Printed on acid-free paper

Printed in the Netherlands.

To the students of the various faculties
of the University of Genoa, Italy,
who enthusiastically participated in courses
in which the content of this book was experimented

TABLE OF CONTENTS

FOREWORD

by Elisabeth Mann Borgese
Founder and Honorary President
International Ocean Institute

Adalberto Vallega has been, for decades, a master and great teacher of integrated coastal management and Mediterranean cooperation. This new book, of an almost encyclopaedic scope, is a most original contribution to the rapidly growing literature on the subject, of equal value to the academic community which will greatly appreciate the theoretical, historic and philosophical underpinning of the work, and to the practitioner, the planner, regulator and manager, who will find in these pages most useful "checklists" for his duties and responsibilities.

Vallega perceives the need for Integrated Coastal Area Management (ICAM) in the broader context of the ongoing third industrial revolution, which he calls the *trans-industrial stage,* in its interaction with climate change.

There have been profound changes in the economies of the industrialized countries. The development of the new High Technologies, including micro-electronics, genetic engineering, new materials, has accelerated the transition from an economic system based primarily on production to one based very largely on services. This, in turn, has facilitated "globalization" of production systems and services, including the financial system, as well as the migration of people The ongoing global "Great People's Migration" is, generally, from the hinterland to the coasts where, already today, over 60 percent of the human population resides, exercising unprecedented pressures on the coastal and marine environment. Clearly, this justifies the current emphasis, at global, regional and national levels, on the need for coastal management.

This increasing pressure on the coastal and marine environment may have more than local effects. It not only threatens human health in densely populated coastal areas as well as the survival of many of the oceans living resources; it may also accelerate a climate change which appears to be in the making although there are as yet many uncertainties both about its natural causes and the interaction between natural and anthropogenic causes. Uncertainty, indeed, is the name of the game: "... the rates of change may be unprecedented.... This approach is invaluable for coastal management since it leads managers and planners to avoid mechanistic and deterministic approaches to the ecosystem," Vallega points out, "To consider the future of the physical environment as evolving non-linearly, and the evolution of the ecosystem as unpredictable, is a very intriguing exercise, not least because it requires setting up management programmes and plans supposing a future that cannot consist of the projections of present trends. However, this exercise is unavoidable: its practice marks a threshold between the conventional and innovative approach to the coastal systems."

This is the broad setting within which Vallega, with truly Latin sense of rigour and architecture, composes a modern system of genuinely integrated coastal area management for sustainable development.

This system comprises both the well-defined physical and the human dimensions. Sustainable development has its ecological, economic, and ethical dimensions.. Each Chapter has its theoretical part, summed up in useful tables and matrices, and illustrated with practical experiences; and each chapter tests its theory on the Mediterranean case study, or examines the Mediterranean situation in the light of this theory. The whole book moves systemically from the physical to the human component and integrates them in the UNCLOS/UNCED legal framework, an emerging institutional or *governance* framework and planning, programming and managing framework

With regard to the physical component it should be noted that Vallega includes a succinct but very clear description of plate tectonic theory, mostly omitted in the literature on coastal management which tends to focus too narrowly on living resources and pollution. And yet, plate tectonics, the genesis of active and passive continental margins has indeed profound effects on coastal management which has to cope with the volcanic phenomena, the tsunamis, the earthquakes which may derive from it.

The UNCLOS/UNCED legal framework is analysed from the point of view of the coastal manager whose tasks and responsibilities, and limitations in each juridical zone —internal waters, territorial sea, contiguous zone, exclusive economic zone and continental shelf—are derived from the legal documents, compiled and integrated in extremely useful tables.

In dealing with the emerging institutional/governance framework, Vallega, in accordance with Agenda 21, stresses the importance of the level of the local, coastal community, where a decision-making system is suggested that should comprise all major users (*horizontal integration*) and provide a forum for joint decision-making with provincial and state authorities (*vertical integration*). Vertical integration, of course, must not stop at the state level, but must continue through the regional level up to the United Nations, in a system that must be as consistent, coherent, comprehensive and participatory as its parts.

At the local level this kind of horizontally and vertically integrated management/governance system is often called "co-management" system, and this may well be one of the most fertile socio-political concepts of our time. Within the broader context of Vallega's work, the local community, thus organized, may assume unprecedented importance: It may become the most effective "countervailing force" against the negative effects of "globalization"—cultural, ethical, economic, or social. It may bind or "integrate" multinational companies into local communities as participants in the common task of enhancing sustainable development. It may enhance the process of democratization, domestically and internationally. One might go as far as to say: this is the emerging shape of democracy in the next century; for democracy, as everything else is bound to change. Quite conceivably, the political-party based democratic system, the "Westminster model," which, emanating from the UK and France, has conquered the world, may not be there to stay for long. Widely corrupted and "commercialized" in the "North," it may not have been the most suitable system for the South to begin with. A community-based democracy may accord better with the historical evolution and the indigenous cultures of developing countries than a political-party based system.

The socio-political map is going through a process of basic reshuffling through two, simultaneous and complementary processes: States are breaking up and/or devolving powers to smaller units, yielding to cultural, linguistic, religious or similar pressures. States are yielding sovereign powers to larger, regional organisations ("economic integration organisations") due to economic, environmental and ecological imperatives. The emergence of local co-management systems for integrated coastal area management may well be a part of this process.

The design of this book was mainly due to the discussions which developed in the context of the International Geographical Union (IGU) and UN organisations. In 1996, during its 28th International Geographical Congress, the General Assembly of IGU resolved to include the programme *Oceans* in the 1996–2000 activity programme. Education in the ocean field, including coastal areas, was regarded as a main component of *Oceans* in a view of contributing to the implementation of interdisciplinary approaches tailored to the guidelines from Agenda 21, adopted by the United Nations Conference on Environment and Development (UNCED, 1992). In 1996, in the context of *Oceans*, IGU and the Intergovernmental Oceanographic Commission (IOC) of UNESCO decided to explore the possibility of establishing cooperation on research and education and, from their initial consultations, education on coastal management was regarded as a core subject on which efforts would effectively converge. Discussions led the Intergovernmental Oceanographic Commission of UNESCO and the International Geographical Union to convene the International Seminar *The role of ocean science and geography in facing ocean managment for the twenty-first century* (Sagres, Portugal, 3–5 September 1998) where the joint programme *Oceans 21—Science for sustainable use of ocean and coastal zones* was adopted. This new and extended programme was intended to absorbe and susbstitute the "Oceans" programme with the aim of carrying out investigations on an interdisciplinary basis, focusing on the interaction between human communities and the coastal and deep-ocean ecosystems, and supporting policy.

In that context the International Conference *Education and Training in Integrated Coastal Area Management. The Mediterranean Prospect* (Genoa, Italy, May 25–29, 1988) was designed in order to provide room for the interchange of discussions involving intergovernmental organisations, educational bodies and experts. That event was convened to celebrate the *1998 International Year of the Ocean*. Its organisation was entrusted to the International Centre for Coastal and Ocean Policy Studies (IC-COPS), a UN non-governmental organisation with the observer status to the Barcelona Convention. The Mediterranean Action Plan (MAP) of the UN Environment Programme (UNEP), and the International Centre for Science and High Technology (ICS) of the UN Industrial Development Organisation (UNIDO), together with the UNESCO as a whole, joined IOC to provide scientific and technical collaboration with the Conference. The European Commission, Directorate General XII (ED DG XII), gave its support. Discussion led the participants to adopt resolutions regarding a project of a master degree in integrated coastal management, the establishment of a network of universities and educational bodies, and that of a network of bodies involved in coastal Geographical Information Systems.

This book was designed and written in this framework of initiatives where the scientific approach of IGU, focusing on the role of deep-ocean and coastal areas within global change, and the programme areas of intergovernmental organisations, rooted on Agenda 21, have been regarded as the main references. The more the book is a basis for discussion in the international arena, the more it will meet the stimuli from which it was conceived and will contribute to implement co-operation and partnership between the scientific milieus, as well as between these and the decision-making centres. Discussion on the book and, *sensu lato*, on the issues of coastal education, may be developed by co-operating with the initiatives carried out by the International Geographical Union, especially with those convened jointly with the Intergovernmental Oceanographic Commission of UNESCO.

Adalberto Vallega
Vice-president of International Geographical Union

ACKNOWLEDGEMENTS

Many people have encouraged me to conceive and write this book. A key role was played by Professors Biliana Cicin-Sain and Robert W. Knecht, co-directors of the Center for the Study of Marine Policy, University of Delaware (Newark, USA). Interestingly, in 1996–1997 their book "Integrated Coastal and Ocean Management. Concepts and practices"(published in 1998), was being written while this book was in the making. Also the experiences and talks with Professor Elisabeth Mann Borgese, Founder and Honorary President of the International Ocean Institute, contributed significantly to my work. Discussions with Prof. Bruno Messerli, President of International Geographical Unions (IGU), on the need to build up educational tools in the framework of the IGU *Oceans* programme with the aim of linking the concepts of global change and sustainable development, had a driving role. In the framework of the Intergovernmental Oceanographic Commission (IOC) Dr Gunnar Kullenberg, at that time serving as the Executive Secretary, had a similar stimulating role.

On the academic side, I should also like to acknowledge Dr Hance D. Smith, Chairman of the IGU Commission on Marine Geography, Professor Pieter G.E.F. Augustinus, Chairman of the IGU Commission on Coastal Systems, and Professor Stephen Olsen, Director of Coastal Resources Center, The University of Rhode Island (Narrangansett, USA), for the opportunity of discussing conceptual and methodological issues and sharing experiences. As regards UN organisations, Prof. Ugo Leone, Science Senior Consultant of UNESCO, helped me to focus on subject areas concerned with the need for professional skills in the intergovernmental organisations. The advice of Dr Lucien Chabason, Co-ordinator of the Mediterranean Action Plan (MAP) of the UN Environment Programme, Dr. Ivica Trumbic, Director of the Priority Actions Programme, Regional Activity Centre (PAP/RAC) of MAP, and Dr. Arsen Pavasovic, former Director of PAP/RAC, led me to focus on case studies using the Mediterranean as the geographical area of reference.

I am grateful to Professor Annalisa Calcagno Maniglio, President of the International Centre for Coastal and Ocean Policy Studies (ICCOPS; Genoa, Italy), a UN non-governmental organisation with the observer status to the Barcelona Convention, and the Executive Committee of this NGO for partly supporting the book. The Department Polis (Department of Urban and Regional and Landscape Planning; University of Genoa, Italy) had the merit of including pilot courses on integrated coastal management in the curricula of the Faculty of Architecture, where the approach presented in the book could be experimented. Moreover, this Department has offered its Internet site to host a homepage and an e-mail round table dedicated to the book aimed at stimulating discussions on its approach.

I should especially like to thank Dr H.D. Smith for reviewing the book. A language review was also provided by Professor Arthur Lomas and Dr James Logan Reynolds, co-operating with the Department Polis. Nevertheless, I would note that any mistakes are my sole responsibility.

Numerous people have given technical collaboration. Significant contributions include Dr. Stefano Belfiore, Secretary of ICCOPS, who edited the book, Dr. Antonella Primi, Department Polis, who assisted in the bibliography and references, and Dr. Patrizia Verteramo, Department Polis, who did the drawings.

Finally, loving thanks are due to my wife, Bruna, and my children, Patrizia and Giorgio, who solicitously helped me to cope with personal impediments during the preparation of the book.

Adalberto Vallega

INTRODUCTION

The Rise of the Coastal Concern

In the early 1990s a scream of anguish was heard all over the world: the population of coastal areas had increased much more than any other areas. At the end of the twentieth century, about two thirds of the world's population live within 60 miles of the coastline and it is expected to continue rising. Moreover, the scenarios of climate change and subsequent sea-level rise have diffused throughout the perception of decision-making systems as well as coastal and island communities. To deal with these combined natural and social changes Agenda 21 (United Nations Conference on Environment and Development, UNCED, 1992) elected to pursue sustainable coastal development by carrying out integrated coastal management. Its Chapter 17 regarded this as a core programme area and enunciated an extended set of sector issues, including capacity building, to implement it on all the scales, from global to local. In its turn, science, although quite unprepared to meet that political impulse, has been in agreement that, in the present historical phase, integrated management is the sole response which mankind might make to coastal change.

Besides the ambiguity of "integrated management", where there has been much speculation since the early 1990s, the undertaking of coastal management in its pursuit of sustainable development has been correctly perceived by the scientific community as an unprecedented binding task. The joint pursuit of the integrity of ecosystems, economic efficiency and social equity, which comprise sustainable development, is hard to undertake in any kind of region and social milieu. However, it is even more in coastal areas because of the complicated web of local ecosystems linking the land, the sea and the near atmosphere, the increasing human pressure and the expanding use of natural and cultural resources.

To ensure some prospect of success for the efforts at integrated coastal management and to disseminate relevant programmes in both developing and developed worlds, education deserves to be regarded as a main component of capacity building. Implementation of efforts in this field is recommended by Agenda 21 and also by intergovernmental documents adopted on the regional (multi-national) scale. On this basis, bearing in mind the long experience accumulated in countries, such as the United States, the United Nations organisations have conducted education courses in the framework of both academic and extra-academic curricula. This experience has made it self-evident that the effectiveness of efforts largely depends on international co-operation on both global and multi-national scales. This recognition has drawn attention to those semi-enclosed and enclosed seas, and other multi-national coastal and island spaces, where co-operation in the field is facilitated by the existence of intergovernmental agreements, such as regional conventions pursuing environmental protection and economic collaboration on sustainable development.

1

The Basis of Coastal Assessment: Epistemology, Logic, Methodology

International co-operation is necessary, but not sufficient, to implement education and training successfully. Inter-disciplinarity is also needed because Agenda 21-based approaches to coastal management require holistic views that only close integration between disciplines is able to provide. Hence an increasingly wide discussion, involving both intergovernmental organisations and scientific bodies, has focused on how inter-disciplinarity should be intended. As is well known, textbooks on coastal management are essentially concerned with the national scale and concentrate on the institutional frameworks and core issues. To complement the panorama this book aims to provide bases for designing and implementing coastal management programmes on the local scale. It intends to show how the coastal manager and planner may design management programmes and plans to operate sustainable coastal development moving from the integrated management principle. In this respect, concepts, logical frameworks, and methods are the subject areas of the approach. The main task is to build up a basis on which to assess the individual, sectoral issues.

To play this role, it was necessary to adopt an explicitly enunciated background thereby taking sides in the present rich and somewhat confusing horizon of conflicting paradigms and hazy concepts. In this respect, sides were taken on three levels, epistemological, logical and methodological.

As regards *epistemology*, the book is based on the postulate that a feedback exists between the holism principle and inter-disciplinarity: the more effective the inter-disciplinarity the more holism is operated; the more holism is operated, the more effective the inter-disciplinarity. To put this feed-back in motion it is necessary to abandon positivist epistemology, influencing Nature's, mainly physical and chemical sciences, and the structuralist approach, still widely influencing social sciences, as they both lead to merely assembling disciplinary approaches. This frustrating conventional situation may be transcended if a mutual epistemological basis is adopted supporting both the scientific arenas. To this end the book considers the epistemology of complexity as a mutual basis for integrating disciplines involved in coastal management and assumes the coastal area as a complex system.

As regards *logic*, to be consistent with the epistemological choice, the background supporting the approach refers to the principles from the general system theory which are noteworthy antipodal to those of the Cartesian logic, supporting both positivist and structuralist approaches. Clearly, the holism principle is not part of Descartes' thought and so there is no reason to adopt this principle as the leading tool for integrated coastal management and, in the meantime, to continue referring to a logical framework, i.e. the Cartesian one, which cannot host it. This contradiction, present in most literature on coastal area management, is tentatively overcome. The results of this operation are submitted to the discussion by scientists and experts involved in coastal research.

As for *methodology*, the consequences from the methodological choice are drawn explicitly. As far as possible, analytical and deductive approaches are not applied and axiomatic-inductive approaches are adopted.

The Setting of the Book

After the presentation of the mentioned background (Chapter 1), the book follows a pathway reflecting the sequence of components of sustainable development as closely as possible. The basic component, i.e. the integrity of the ecosystem, supports Chapters 2 and 3 where the assessment of abiotic elements and the trophic organisation of the coastal area useful for carrying out management programmes, is presented. The second and third components of sustainable development (economic efficiency and social equity) are dealt with in the following chapters. First the legal, administrative and jurisdictional frameworks with which the coastal manager is concerned are presented (Chapter 4). Then attention shifts to the coastal area economic organisation proposing an approach tailored to the holism principle (Chapter 5). The joint consideration of the subjects discussed in Chapters 2 to 5 leads to dealing with a classic issue of coastal management: how to delimit the coastal area (Chapter 6). At this point a comprehensive basis for understanding the nature and role of coastal management may be regarded as set up. So attention moves to a range of key questions: how the uses of natural, human and cultural resources may be assessed through the concept of coastal use structure (Chapter 7), and conflicts between uses may be approached in order to prevent and mitigate them (Chapter 8); how the role of the decision-makers may be assessed and its optimum setting may be prefigured (Chapter 9). The last chapter is concerned with the design of the coastal management programme. A concise panorama of the approaches from inter-governmental organisations and scientific literature introduces the presentation of a framework design of the integrated management process.

The Mediterranean as the Case Study

The book selects the Mediterranean as the geographical area to which to refer when concepts and methods are presented and discussed. There were many reasons for not presenting case studies from many other parts of the world and for concentrating attention on the Mediterranean. The continental coastal areas, as well as small islands, of this basin are characterised by great physical complexity and biological diversity and have been subject to great human pressure. Moreover the marine space was the first semi-enclosed sea to benefit from an Action Plan (1975) undertaken within the framework of the UNEP Regional Seas Programme. There is a regional convention, the Barcelona Convention (1976), initially conceived to pursue the protection of the Mediterranean marine environment which has recently (1995) been implemented in the objective framework with the inclusion of integrated coastal management. The Mediterranean Action Plan (MAP) has carried out pilot programmes on coastal management based on co-operation between UNEP, national and local organisations. The guidelines provided by UNEP (1995) on integrated coastal management were enunciated especially bearing in mind the results and lessons from coastal management in this basin. Finally, the experiences in the Mediterranean could contribute to coastal management programmes in other regional seas.

The Setting of the Chapters

Most figures in the book aim to illustrate webs of concepts, principles and data to help clarity of the text and provide means for applying its approach. References and the index are divided into sections, respectively concerning the general aspects of coastal management and the Mediterranean.

Each chapter setting presents the subjects to be dealt with, and usually consists of three parts. The first presents concepts, logical and methodological approaches and discusses the contributions from inter-governmental organisations and scientific literature. The second focuses on the Mediterranean applying concepts and methods to this geographical area. The final part proposes an overview of subjects, issues, recommendations or guidelines to be put into the agenda of coastal managers keen to operate integrated management. Where needed, terminological and topical *mises au point* are presented in boxes.

CHAPTER 1

THE SUBJECT

The objective of this Chapter is to give a basic view of the subject areas of the book. The view will be concerned with the conceptual framework of, and approaches to, integrated coastal management. The reasons why the Mediterranean is a significant case study will be presented. The following questions will be discussed:

Questions	Figures	Tables
Why coastal management?	1.1	
Why does coastal management imply taking into account epistemological and methodological important issues?		1.3
How has coastal management evolved?	1.1	1.2
How may sustainable coastal development be achieved?		1.1
How may integrated coastal management be achieved?		1.4
What relationship exists between the concept of integrated coastal management and the system-based logical background?		1.3
How should "coastal area" be intended?	1.2	
How does the concept of "coastal area" differ from those of "coastal zone" and "coastal system"?	1.2	
Why is it useful to consider the Mediterranean as a core area study?	1.4	1.5, 1.6, 1.7
How meaningful is the Mediterranean experience in the field of coastal management?	1.3, 1.5	1.6, 1.8

1.1. The Reference Basis: Agenda 21

The basic criteria through which to approach integrated coastal management (ICM) were enunciated by Chapter 17 of Agenda 21, adopted by the *United Nations Conference on Environment and Development* (UNCED, Rio de Janeiro, 1992). Chapter 17 includes seven programme areas (Table 1.1), the first of which is entitled *Integrated management and sustainable development of coastal and marine areas, including exclusive economic zones*. This title is substantial because it links the concept of ICM with the principle of sustainable development and leads to the pivotal statement supporting the approach to the subject: *integrated management is instrumental for sustainable development*. As a result, "coastal States commit themselves to integrated management and sustainable development of coastal areas and the marine environment under their national jurisdiction" (Agenda 21, Paragraph 17.4). Among the main consequences of this approach is the geographical coverage of ICM. It consists of programmes and plans jointly and contextually involving both coastal land and marine areas (Barcena 1992). It is also rich in a cultural and political sense due to the fact that, for the first time in its history, mankind decided to pursue a final objective (sus-

5

tainable development) throughout the world by adopting a well designed tool (integrated management).

TABLE 1.1.   Agenda 21, Chapter 17. Programme areas

Programme areas	Paragraphs
A. Integrated management and sustainable development of coastal and marine areas, including exclusive economic zones	17.3 to 17.17
B. Marine environmental protection	17.18 to 17.43
C . Sustainable use and conservation of marine living resources of the high seas	17.44 to 17.69
D . Sustainable use and conservation of marine living resources under national jurisdiction	17.70 to 17.96
E. Addressing critical uncertainties for the management of the marine environment and climate change	17.97 to 17.115
F. Strengthening international, including regional, cooperation and coordination	17.116 to 17.123
G. Sustainable development of small islands	17.124 to 17.137

At least four motivations sustain the approach in Agenda 21.

First, *human pressure*. "The coastal area contains diverse and productive habitats important for human settlements, development and local subsistence. More than half the world's population lives within 60 km of the shoreline, and this could rise to three quarters by the year 2020" (Agenda 21, Paragraph 17.3; see also Edwards 1989). As a result, an increasing number of coastal areas have been affected by critical interaction between (i) the demand for space and natural resources which has been brought about by both the growing local population and immigration of peoples from inland, and (ii) what is available in coastal areas in terms of sites for settlements, freshwater, energy sources and other resources.

Secondly, *global environmental change*. Global atmospheric warming is expected to deeply influence marine erosion processes by accelerating sea-level rise, as well as inducing changes in freshwater circulation, with subsequent impacts on marine waters. As a result, coastal areas may become that part of the Earth mostly involved in the consequences of global environmental change which was initiated in the mid-nineteenth century and which has accelerated during the second half of the twentieth century.

Third, the *abundance of resources*. The marine biomass overlying the continental shelf provides vast living resources, the bed and subsoil are rich in oil and gas, coastal agriculture is characterised by high productivity, especially in inter-tropical and temperate latitudes, and aquaculture in coastal waters is an essential resource for future generations. Also the cultural heritage is an important resource since many ancient civilisations were born and have flourished in coastal and island regions.

Fourth: *underdevelopment*. "Many of the world's poor are crowded in coastal areas" (Agenda 21, Paragraph 17.4). This condition can be one of the most binding issues for future generations. The more the destruction of biomass on the inter-tropical latitudes where less developed countries are concentrated, propagates because of the

growing human pressure, tourism and other factors, the more the availability of natural coastal resources is jeopardised (Golberg 1994). As a result, the coastal areas and
islands of these countries are part of the world where underdevelopment is expected to
display its most critical features.

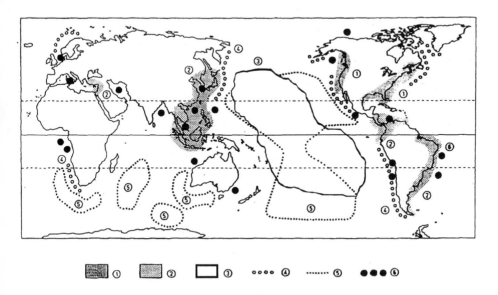

Figure 1.1. World geographical view of coastal and ocean management framework: 1) coastal areas
where all the States are carrying out integrated coastal area management giving shape to mature coastal
regions; 2) id., where 40–60 percent of States are carrying out coastal area management programmes, so
moving towards the creation of coastal regions; 3) ocean spaces where the building up of deep-ocean
regions is expected; 4) zooplankton abundance areas (more than 500 mg/cu/m); 5) deep-ocean areas
withmanganese nodule deposits; 6) major oil and gas exploitation ocean areas.

As a result, coastal management is expected to become one of the most demanding
tasks for future generations. In many respects, the UNCED preparatory materials emphasised that the key role of the ocean in the Earth's ecosystem is largely due to the
fact that the saltwater surface is 70 per cent of that of the Planet— 2.4 water sq km to
each land sq km. Hence, the *Blue Planet* denomination attributed to the Earth has acquired a double sense, indicating how much climatic change may interact with the
ocean, and how extensively the ocean may participate in the future world organisation. Moreover, since the coastal interface between land and the ocean is 603,000 km
length, equal to 15 times the Equator, the Earth is also a *Coastal Planet*. Coastal and
island states—including territories with limited sovereignty, such as those adminis-
tered by the United Nations, and those committed to trusteeship—are totally 166, of
which 60 (36 per cent) are inland and archipelagic states encompassing about 30 per
cent of the world coastline extent.

1.2.   Evolution of Coastal Management

The perception of the prominence of coastal area management arose in the 1980s and, in general, was stimulated by two basic needs: on the one hand, that of facing coastal erosion and subsequent environmental degradation, including natural disasters; on the other hand, that of pursuing economic development through the implementation of core sectors, such as seaports and recreational facilities (Couper 1992). In that process two key-words, namely, environment and development, were catalytic. The anniversary of the political approach to coastal management dates back to the adoption of the *Coastal Zone Management Act* (CZMA, 1972) by the United States. That legal instrument defined objectives, enunciated concepts and principles, and provided guidelines on which, for many years, national governments and subnational decision-making centres would have based their policies for coastal management (Barston 1994). Furthermore, bases for undertaking and implementing coastal management programmes in developing countries were provided. That step stimulated other countries, initially European (Ballinger, Smith H.D. & Warren 1994), to adopt measures to protect their coastal areas and sustain their economic development (Vallejo 1988: 205–6; Smith H.D. 1994). In 1972 the *United Nations Conference on the Human Environment* (Stockholm) was convened. Although its materials and resolutions were not specifically concerned with coastal management, the impetus that they gave to the design and diffusion of environmentally-sound policies have also deeply influenced the criteria and methods used to approach this issue (Peet 1992; Smith H.D. 1992).

1.2.1. TAKE OFF

As a result of this wide range of impulses, in the 1970s the take off phase of coastal management took shape. The UN had a leading role in encouraging states, with special consideration for developing countries, to undertake initiatives in the field. In 1977 a report by the UN Secretary-General (UN Document E/5971; reproduced in Ocean Yearbook 1, 354) pointed out that the new economic frontier marking the coastal regions was the result of two core social objectives, *development* and *management*, and consequently it was based on a hybrid concept "combining elements of regional analysis and environmental management". That approach emphasised for the first time that the complexity of coastal management largely depends on the need to adopt concepts and methodologies provided by regional science. Initially the approach was *multi-disciplinary*: it associated natural and social sciences involving the former much more than the latter. Later on, the approach was symptomatically *inter-disciplinary* because it was increasingly characterised by efforts to integrate conceptual and methodological contributions from different disciplines and sectoral views. Integration was perceived as essential as far as it was self-evident that environmentally-sound and development-aimed strategies ought to be carried out jointly. That approach made the scientific community and decision makers aware of an intriguing issue: environmental management demands minimising the social impact on the ecosystem while development aims at maximising resource uses.

At the beginning of the 1980s the UN (United Nations 1982: 4–8) pointed out that the maximisation of both economic efficiency and environmental quality was not contextually pursuable. A correct approach ought to respond to the question of to what extent it should be guaranteed that fixed thresholds of environmental impact would not be exceeded while expected economic objectives were reached. That statement was perceived by coastal managers as the basis of the *economic principle*, according to which any coastal management pattern has a cost: "one cannot get something for nothing" (*Ib.*: 5). However, society requires objectives for coastal management to be designed by accurately assessing the environmental impacts and by adopting measures apt to protect the ecosystem from structural changes. As a consequence, the economic principle ought to be complemented by the *ecological principle* (*Ib.*: 17–8) stating that the coastal ecosystem is an organism whose components are strictly tied to each other creating food chains and storing energy. To be consistent with such a perception of Nature, coastal management should have been regarded as a set of measures to exploit food chains and use energy without interfering with the organisation of the ecosystem.

1.2.2. DRIVE TO MATURITY

During the 1980s coastal management reached its pre-maturity phase. Evolution was due not only to the diffusion of coastal management programmes but also to other factors, *inter alia* the fast and strong development of conceptual tools, the intense debate within the scientific community, the increasingly close interaction between scientific bodies and decision-making systems, and the leading role undertaken by the UN (United Nations 1982). The visible effects of that increasing attention to the field and the implementation of both the scientific and political approaches were remarkable. Coastal management programmes progressed from the approach to a single or a few uses, as well as the single environmental impacts, which marked the take off phase (the 1970s), to multiple use-based programmes. An increasing number of experts in social sciences, mainly in decision-making processes, and coastal and marine biologists joined geomorphologists who had played a leading role during the late 1960s and the 1970s (Vallejo 1992). The coastal area was imagined as extending much more seawards than had been considered the case in the past. The UN system, mainly through the UN Environment Programme (UNEP) and the Food and Agriculture Organization (FAO) and using resources from the World Bank and the UN Development Programme (UNDP), became the main component of that process.

The conceptual distance covered from the CZMA has been correctly emphasised by Olsen (1995: 4): "We have progressed since the early 1970s from coastal zone management (CZM), which recognized some of the more obvious development 'mistakes' and social conflicts along the shorefront and responded with various forms of regulation to ICM that attempts to address the assumptions and policies underlying the development process and to experiment with new approaches to governance at the community level and within the agencies of central government". However, such a conceptual improvement would not have been accomplished if three other factors had not been taken into account.

TABLE 1.2.   Stage-based model of coastal management

Stage	Objective	Coastal uses under management	Geographical coverage
Late Sixties: pre-take off	Use management facing a single environmental issue socially perceived as important.	One or a few uses (e.g., seaports and recreational uses).	The shoreline.
Seventies: take off	Use management and environmental protection.	A few uses (e.g., seaports, manufacturing plants, recreational uses and fishing).	• The shoreline; • A coastal zone delimited according to arbitrary criteria; • Id., according to administrative criteria.
Eighties: drive to maturity	Use management and environmental protection.	As an orientation, the whole range of uses.	As above, seawards tending to manage national jurisdictional zones.
Nineties: maturity	Integrated management.	Comprehensive use management. The management of the coastal ecosystem.	Coverage delimited: • landward, according to various criteria; • seaward, with reference to the extent of the widest national jurisdictional zone.

First, *global change*. The international research programmes on climate change and subsequent changes in biogeochemical cycles, on the one hand, and social change, on the other, made it clear that special attention should have been given to coastal areas. In particular, investigations of sea-level variations dependent on atmospheric warming made it possible to design scenarios of coastal erosion and its impacts on the national and subnational scales. The rapid growth in assessing these physical processes and consequences for the coastal ecosystems, man-made facilities and settlement organisation made it self-evident that *ad hoc* coastal management programmes were needed.

Secondly, *sustainable development*. During the 1980s the rise in theoretical approaches within the UNESCO framework, mainly in the context of the Man and Biosphere (MAB) programme, led to defining the concept of sustainable development. It was intended as a system of goals consisting of: (i) the integrity of the ecosystem, (ii) economic efficiency, and (iii) social equity including the rights of future generations. The World Commission on Environment and Development (1984–1987)[1] moved from that speculative basis to propose sustainable development as a political paradigm able to guarantee the contextual pursuit of the environmental and development goals on any scale, and to trace a new course to reduce the gap between developed and developing countries. This view of the Commission was agreed by the UN General Assembly which decided (Resolution 44/228, 22 December 1989)[2] to convene the UN Conference on Environment and Development (UNCED, Rio de Janeiro, 1992). That political process enabled integrated management to be assumed as the goal throughout the world.

TABLE 1.3. Conjunctive *versus* disjunctive logic. Application to coastal management

Disjunctive logic	Conjunctive logic
Cartesian thought	Theory of the general system
Principles	
Evidence	Pertinence
Describing only those elements of the coastal area which are clear.	Describing those elements of the coastal area which are perceived as essential to management.
Reductionism	Holism
Desegregating the components of the coastal area, then describing the individual elements, separately.	Describing the coastal area as a unique system interacting with its external environment. Considering detailed knowledge of the coastal structure as not essential.
Causality	Teleology. [3]
Moving from the simplest elements of the coastal area towards the most complicated. Supposing that the role of the elements is regulated by cause-effect relationship.	Leaving out the existence of a cause-effect relationship between the elements of the coastal area. Focusing on feed-back and circular relationships. Considering the evolution of the coastal system with reference to its project, i.e. the goals which are pursued through its organisation.
Exhaustiveness	Aggregativeness
Assessing all the elements of the coastal area in detail. Ensuring that nothing has been left out. Postulating that, only where these conditions are met, the subsequent knowledge is objective.	Selecting those elements which pertain to coastal area management, i.e. the goals and projects of the coastal system, and leaving out the rest. Being aware that knowledge is relativist and partisan per se.

Third, *epistemological changes*. During the 1980s, mainly within the UNESCO framework, intense epistemological discussion developed involving important components of the scientific community. It concerned the innovative role of the theory of complexity, also called theory of the complex system, *vis-à-vis* the conventional epistemological background, more or less rooted on positivism and structuralism, and using disjunctive and deductive logical approaches (Table 1.3). There is no room here to discuss it. It is sufficient to note that the conventional approach was based on the exhaustive description of reality and its sectoral analysis (principles of exhaustiveness and reductionism). As regards the coastal area, that approach would have led to pseudo-objective and detailed analyses which would not have guaranteed achieving results consistent with the need to carry out ICM. On the contrary, the theory of complexity has led to global views of reality by selecting and focusing on those elements closely referring to the objective of assessment and basing investigations on them. In so doing, scientists were able to provide goal-oriented views tailored to the guidelines of Agenda 21 which, as regards the coastal area, recommend working out holistic and management-oriented approaches.

In conclusion, as far as coastal systems are concerned, the three paths (global change, sustainable development and the epistemology of complexity) have converged on the same goal: ICM. Research on global change provided the awareness of the need to harmonise human action with the evolution of Nature. The sustainable development principle designed the final goal to which the specific goals, pursued in the single coastal areas, should be correlated. The theory of complexity made it possible to build methodological frameworks tailored to the concept of integrated management.

1.3.   Sustainable Coastal Development

As has been mentioned, ICM was considered in Agenda 21 as a tool to pursue sustainable coastal development. It follows that, before considering the idea of ICM it is necessary to present the concept of sustainable development. The conventions and Agenda 21, as well as other documents adopted by UNCED, do not provide a definition of sustainable development. The *Rio Declaration*, conceptually supporting all the materials adopted by UNCED, includes some statements on how sustainable development should be intended but does not provide a definition. The report of the World Commission on Environment and Development (1987: 43) states: "Sustainable development is development that meets the needs of the present without compromising the ability of future generations to meet their own needs. It contains within it two key concepts: the concept of 'needs', in particular the essential needs of the world's poor, to which overriding priority should be given; and the idea of limitations imposed by the state of technology and social organisation on the environment's ability to meet present and future needs (...). Development involves a progressive transformation of economy and society. A development path that is sustainable in a physical sense could theoretically be pursued even in a rigid social and political setting. But physical sustainability cannot be secured unless development policies pay attention to such considerations as changes in access to resources and in the distribution of costs and benefits".

This statement makes it clear that sustainable development is a system of goals having the role of a political paradigm in the sense that it derives from a view of how the Earth's ecosystem and the world should be governed by present and future generations. This system consists of three components (Young 1992: 19–20):
a) the integrity of the ecosystem;
b) the efficiency of the economy;
c) social equity, including the rights of future generations;
and it is liable to produce vast consequences for coastal management.[4]

The Role of the Ecosystem
Between the two UN conferences—i.e. the 1972 Conference on the Human Environment and the 1992 Conference on Environment and Development—a profound change took place. The former event inaugurated environmentally-sound policy, basically aimed at preventing and mitigating pollution, thereby dealing with the abiotic compo-

nents of the ecosystem and involving, solely or mainly, physical and chemical disciplines. The latter event focused on the ecosystem as a whole and attributed special attention to the protection of its biotic components. Biology was associated with physical and chemical sciences and, thanks to the impulses by the Convention on Biodiversity (1992), became a chief component of the scientific approach to the environment.

Economic Efficiency versus Productivity

The concept of sustainable development has two main consequences. First, a holistic view of resources is needed since natural, viz. biotic and abiotic, as well as human and cultural resources are to be regarded as essential to correctly pursue economic and social development. This approach is much broader than the conventional one, because it attributes basic relevance to the availability of natural resources, both at the present time and in the future. Secondly, the conventional approach is based on the idea of productivity since development is regarded as the maximum value of the ratio between costs and gross product. In contrast, the innovative approach is based on the idea of efficiency, since development is intended as fundamentally consisting of the minimum use of non-renewable natural resources, the optimum use of renewable ones, the guarantee of optimum quality of life and the conservation of the cultural heritage. In short, efficiency is the result of the internalisation of the ecosystem, the cultural heritage and social equity in the political equations. A much more complicated policy than the conventional one thus arises.

The Role of Ethics

The pursuit of social equity leads policy to focus not only on the human rights, as they were defined by all the solemn declarations more or less inspired by the Enlightenment, but also on the right of access to the ecosystem and cultural heritage from all sectors of the human coastal community. This innovation has acquired growing political importance as ideologies, especially Marxism, are no longer influential and meta-ideological views, basically rooted in ethical values, have gained social justification and have been regarded as new references for policy. Sustainable development is a product of this cultural change.

The Convergence of Three Theories

The adoption of sustainable development as a meta-ideological paradigm implies politics being sustained by the integration of three theories (Young 1992: 19–20):

a) the *economic theory*, which is stimulated to focus upon the economic efficiency and resource rights;

b) the *ecological theory*, which is required to provide conceptual and methodological tools useful for guaranteeing the efficiency of the ecosystem and optimising the use of living resources;

c) the *equity theory*, which is required to design concepts able to innovate policy.

As was stated by the World Commission on Environment and Development, "in its broadest sense, the strategy for sustainable development aims to promote harmony among human beings and between humanity and nature". Nevertheless, this could be a mere and utopian statement if the political impacts brought about by its application was not correctly perceived. According to the Commission (1987: 63) the pursuit of sustainable development needs, *inter alia*:

- "a political system that secures effective citizen participation in decision making;
- an economic system that is able to generate surpluses and technical knowledge on a self-reliant and sustained basis;
- a social system that provides for solutions for the tensions arising from disharmonious development;
- a production system that respects the obligation to preserve the ecological base for development;
- an administrative system that is flexible and has the capacity for self-correction".

1.4. Integrated Coastal Management

The epistemological nature of sustainable development, jointly with the political consequences that it may bring about, are the bases on which to design the concept of ICM. As has been mentioned, Agenda 21, Chapter 17, does not define "integrated management". It only affirms that States are committed to integrated management and sustainable development and must undertake actions necessary to deal with this task (Paragraph 17.3). As a result, the literature has extensively discussed how "integrated coastal management" ought to be intended in conformity with the content of Chapter 17 and the approach which was provided through the four meetings of the UNCED Preparatory Committee.[5] "Integration" derives from the Latin *integru(m),* which means "entire", "undivided". Hence "integrated" has assumed the meaning of "composed or separated parts united together to form a more complete, harmonious, or co-ordinated entity" (Webster's Dictionary).

This justifies believing that, in order to design how *integrated* coastal management should be conceived, one should start from the consideration of the *objective* of management. Although the above mentioned statement (Paragraph 17.3) is not very clear, it is self-evident that, by focusing on the objective, management is basically integrated where plans, programmes and measures are thought of in such a way as to optimise sustainable development (Cicin-Sain, Knecht & Fisk 1995). This implies the content of management—in terms of resource use patterns, organisation of decision-making centres, protection and conservation of the environment and cultural heritage, social participation, and others— being designed in such a way as to optimise efforts to pursue sustainable development. In short, *coastal management is integrated where its specific goals are strictly correlated with the pursuit of the integrity of the ecosystem, economic efficiency and social equity.* Each coastal management programme has its own range of specific objectives depending on the crucial needs to be dealt with—such as coastal erosion or the protection of archaeological remains—and the prospects of development to be realised, such as those regarding tourism or seaports. Correlation is required between these specific objectives and the final, canonical, objective consisting of sustainable development. For example, where tourism is the specific objective, this sector should be organised in such a way as to become a means to protect the local ecosystem and cultural heritage and should ensure the local population benefits from both these patrimonies. Hence the first principle of integrated management is *to maximise the consistency between the specific, local, and final goals*.

Since sustainable development implies the coastal ecosystem being the primary and essential resource for development, coastal management is "integrated" where the ecosystem as a whole is embraced by the programme and subsequent measures. To be explicit, coastal management programmes should cover a given ecosystem, or a set of contiguous ecosystems, and should be referred to all the components, abiotic and biotic, of the ecosystem. As an example, integration does not occur where a Mediterranean bush area is concerned with two or more coastal programmes, each of them covering a part of this space and providing distinct measures. Hence the second principle of integrated management: *to provide a holistic and conservation-aimed management of the individual ecosystem, or a group of contiguous ecosystems*.

As has been mentioned, coastal management programmes are often perceived by the local community and adopted by local decision-making centres with the aim of dealing with a single issue or developing a single economic sector. Management is integrated where the single issue is considered jointly with the other existing issues. For example, if coastal erosion is the main issue of a Mediterranean lowland coastal area, this issue should be tackled not only with the aim of defining man-made coastal structures but also bearing in mind the need to guarantee also the conservation of those components of cultural heritage—such as archaeological remains and historical settlements—jeopardised by coastal dynamics. In short, management is integrated where the development of the single sector is pursued in such a way as not to cause conflicts with other sectors (e.g., seaport development *versus* local marine reserves). Hence the third principle of integrated management: *to design proper cross-issue and cross-sector programmes*.

Correctly Agenda 21, Chapter 21, considers the role of *decision-making systems* as focal. Integrated management is effectively undertaken where all the concerned decision-making centres have the same awareness of the need for coastal management and they adopt co-ordinated actions (Vallejo 1989 and 1992). Co-ordination is needed between decision-making centres involved in different sectors (man-made structures, seaports, fisheries, etc.), as well as between centres pertaining to different decision levels, namely, local authorities, governmental and inter-governmental organisations. Hence the fourth principle of integrated management: *to benefit from inter-sector and inter-rank co-ordination between decision-making centres*.

The Role of ICM According to FAO
The approach by FAO led to designing the role of ICM in the following terms.
  ICM "is concerned with both the conservation and the use of natural resources which include:
- non-renewable natural resources such as minerals and hydrocarbons;
- renewable natural resources such as fresh water, fish and shell fish stocks, kelp beds and mangroves; and
- environmental 'services' and amenities provided by coastal ecosystems including the capacity of the sea to assimilate waste, protection from natural hazards, the maintenance of biodiversity, aesthetic and recreational amenities and waterways and ports.

It is concerned with the organisation and coordination of institutional and other human resources for the purpose of attaining long term strategic objectives.

It is concerned with managing many competing interests and resolving conflicts in relation to the protection, use and exploitation of coastal resources.

It requires an approach which is integrated: across the land/water interface, institutionally (both vertically though the different tiers of government, and horizontally across different sectors of activity), and across different disciplines. It should be noted that integration means more than coordination—it embodies the idea of unification into a comprehensive whole— and that ICAM generally supplements rather than replaces sectoral management.

It is based on management of an area or zone rather than of a particular sector or use".
(From Boealart-Suominen & Cullinan [1994: 8–9]).

Finally, integration is concerned with the *decision-making processes*. In order to optimise the pursuit of sustainable coastal development, the individual measures adopted by decision-making centres ought not to be incompatible with others and, if possible, should interact positively. For instance, the regulations on traffic lanes adopted by the local seaport authority should be co-ordinated with the regulations on fisheries adopted by the fisheries authority. Co-ordination is also needed over time since the time sequences of provisions by various centres should be correlated. For example, the measures for creating a marine park should be harmonised with those regarding urban planning in the neighbouring areas. Hence the fifth principle of integrated management: *to provide co-ordination in content and time between decision-making processes*.

The concept of *external environment* has an essential role. On the one hand, the coastal ecosystem interacts with the surrounding ecosystems and the near atmosphere. As a result, it is subject to many inputs generated by its external environment and, in its turn, impacts upon the contiguous ecosystems with which it is linked. On the other hand, coastal settlements and human activities are influenced by factors acting, and processes developing, in other regions and countries, such as market trends, political conditions and inter-regional and international migrations. In its turn, the coastal community is more or less able to address impulses to other communities and, so doing, to establish interaction with its economic, political and social external environment.

These interactions are closely relevant to coastal management. Their assessment is so relevant as to justify stating that the criteria and principles through which to manage the coastal area ought to be considered as having moved from the understanding of inputs from the external environment. As an example, the protection of the ecosystem requires the pollutants transferred via atmospheric circulation from external cities and industries being assessed. Hence the sixth principle of integrated management: *to co-ordinate the management of the coastal area with that of surrounding land and marine areas*.

To conform to the language adopted by both Agenda 21 and the documents from UNEP, "coastal management" (CM) and "integrated coastal management" (ICM) will be used in this book respectively to relate the subject generally and to refer to coastal management as it was defined by Agenda 21 and subsequent materials. "Coastal area management" (CAM) and "integrated coastal area management" (ICAM) will be used referring, respectively in a general and specific sense, to an individual space with which management is concerned. For example, where the role of decision-making systems is considered *per se*, the term "coastal management" will be used. Where discussion refers to the specific decision-making system operating in a given area, it will be referred to "coastal area management".

TABLE 1.4. Integrated coastal management. Concepts and definitions

Sources	Definitions
	Inter-governmental organisations
FAO, Clark (1992: 7).	"The overall objective of an integrated management programme, like ICZM, is to provide for the best long-term and sustainable use of coastal natural resources and for perpetual maintenance of the most natural environment".
Intergovernmental Panel on Climate Change (1994: 25).	"The most appropriate process to address current and long-term coastal management issues, including habitat loss, degradation of water quality, changes in hydrological cycles, depletion of coastal resources, and adaptation to sea level rise and other impacts of global climate change".
UNEP (1995: 16).	"An adaptive process of resource management for environmentally sustainable development in coastal areas. It is not a substitute for sectoral planning, but focuses on the linkages between sectoral activities to achieve more comprehensive goals".
	Science
Sorensen and McCreary (1990: 17).	"A dynamic process in which a coordinated strategy is developed and implemented for the allocation of environmental, socio-cultural, and institutional resources to achieve the conservation and sustainable multiple use of the coastal zone (from the 'Coastal Area Management and Planning Network, 1989'".
Scura, Chua, Pido and Paw (1992: 22).	"(...) to allow multisectoral development to progress with the fewest unintended setbacks and the least possible imposition of long-run social costs"
Cicin-Sain (1993: 29).	"A continuous and dynamic process that recognizes the distinctive character of the coastal zone—itself a valuable resource—for current and future generations".
Kenchington and Crawford (1993: 111).	"The integration of environmental protection goals into economic and technical decision-making process".
Vallega (1993: 155).	"A holistic approach, in which the ecosystem as a whole (all the biotic and abiotic components) and all kinds of coastal use, as well as all use—use and use—ecosystem relationships, are included".

1.5.   The Coastal Zone and Coastal Area

An important component of the conceptual framework which is needed to deal with ICM is the trio consisting of the concepts of (i) coastal zone, (ii) coastal area, and (iii) coastal system.[6] The literature has made efforts to define these concepts and, in particular, to highlight the differences between them. Comparing the "coastal zone" and "coastal area" concepts, it should be emphasised that some authors consider these, *de facto*, as synonymous, while other authors sustain that they have two distinct meanings. *Inter alia*, this approach is supported by Sorensen and McCreary (1990: 9), who assert that "coastal area" has a the generic meaning of "a geographic space that has not been defined as a zone. In other words, in coastal areas the inland and ocean boundaries to the zone have not been set or approximated". On the contrary, the concept of "coastal zone" is specific and closely concerned with the interactions between human communities and the environment. "The coastal zone—they state—is the band of dry land and adjacent ocean space (water and submerged land) in which land ecology and use directly affect ocean space ecology, and *vice versa*. The coastal zone is a band of variable width which borders the continents, the inland seas (...). *Functionally*, it is the broad interface between land and water where production, consumption, and exchange processes occur at high rates of intensity. *Ecologically*, it is an area of dynamic biogeochemical activity but with limited capacity for supporting various forms of human use. *Geographically*, the landward boundary of the coastal zone is necessarily vague (..)" (*Ib.*: 5; italics mine).

The *Land—Ocean Interactions in the Coastal Zone* (LOICZ), a core project of the International Geosphere—Biosphere Programme (IGBP), uses "coastal zone" and considers it as the area extending "from the coastal plains to the outer edges of the continental shelves, approximately matching the region that has been alternately flooded and exposed during the sea level fluctuations of the late-Quaternary period" (IGBP, Report No. 25, 1993, 9).

Scura, Chua, Pido and Paw (1992: 16-7) speak of "coastal areas" and define them as those areas that "geographically (...) form the interface between land and sea, the complex physical and biological processes played out there testifying to the close terrestrial—aquatic links. Ecologically, coastal areas contain a number of critical terrestrial and aquatic habitats, which comprise unique coastal ecosystems, containing a valuable assortment of natural resources. These interrelated ecosystems are also closely linked with the socioeconomic systems to form resource systems. Resource systems can be conceptualised as encompassing the interactions between biological, terrestrial and marine environments and human activities, including the governing institutional and organizational arrangements".

As regards the institutional literature it is useful to note that UNEP (1995: 51-2) adopted the point of view from the United States. The coastal zone was substantially regarded as that part of the coastal area concerned with a coastal management programme. In the meantime a set of definitions were adopted to drive planners and managers.

The Concept of Coastal Area According to UNEP
"'Coastal area' is a notion which is geographically broader than the coastal zone, the borders of which require a less strict definition. This notion indicates that there is a national or sub-national recognition that a distinct transitional environment exists between the ocean and terrestrial domains. This notion is of extreme importance for ICAM. Many processes, be they environmental, demographic, economic or social, actually take place within the boundaries of the coastal area, with their extreme manifestations being most visible in the area of the coastal zone.
    A graphical presentation of the elements of the coastal zone and the coastal area is given in the figure below (See Figure 1.2) "Ocean waters" cover a narrow near-shore sea belt, its width varying from

one country to another (in the United States this belt is approximately 3 miles wide). "Intertidal area" is the area between the lowest tide line and the shoreline (the landward extent of the tidal influence), including estuaries and coastal wetlands. "Coastline" is the contact line dividing the land from the water bodies. It usually coincides with the line marking the landward extent of tidal influence. "Oceanfront or shorelands area" is part of the land up to the highest line of tidal influence. This is a relatively narrow belt, with its inner borders usually reaching the first coastal road or encompassing the areas reserved for the public access to the coast, protection of sensitive habitats, etc. This belt is rarely wider than 1,000 m. "Coastal uplands" are defined as an area of the interior between the shorelands and, most frequently, the highest peak of the closest mountain range. Sometimes, the depth of the belt is limited (in the United States the limit is 5 miles). "Inland" may be any area outside the aforementioned belts. However, it should not be considered as an altogether unimportant zone, since many processes affecting the state of the coastal zone originate in that area. Generally speaking, that coastal waters, intertidal area, coastline, shoreland area and coastal uplands are the elements of the coastal zone. In addition to the coastal zone elements, the coastal area contains the ocean waters and the inland area".
(From UNEP [1995: 51–2]).

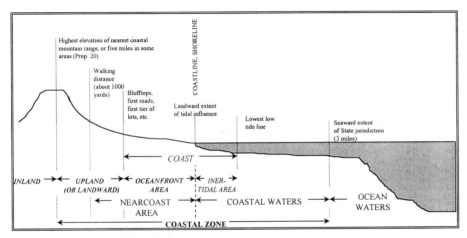

Exaggerated vertical and horizontal scales

Figure 1.2.   The coastal area and coastal zone, according to UNEP (1995, 52).

To conclude this subject, it is worth noting that, fundamentally, the concept of "coastal zone" comes from physical sciences, such as geology and geomorphology. It has been applied to coastal management basically because, in its initial phase, this subject was deeply influenced by these disciplines (Table 1.2). As an example, those research projects, such as LOICZ (UNESCO/ICSU) and ELOISE (EU), which were initiated by scientists from these fields use the term "coastal zone". Instead, "coastal area" is employed more by scientists from social disciplines. *In this book "coastal zone" and "coastal area" will be considered as synonymous and the latter term will be used with the meaning of a land-marine space actually or potentially involved in management programmes.*

1.6. Coastal System and Complex Coastal System

Finally, the theoretical background is concerned with the concept of coastal system. In a general sense, a system is thought of as set of elements, linked to each other by means of an organisational pattern designed to pursue an objective. It may be imagined as an organism, evolving over time, passing through adjustments and structural changes, and so doing moving towards a target. The web of elements of which the system consists, jointly with the web of the relationships between elements, forms the structure. As a consequence, the system could be regarded as a structure which, by force of its organisation, moves towards the achievement of an objective (Vallega 1992: Chapter 1; Van der Weide 1993 and 1996).

The coastal system is a *special system*, basically because it is bi-modular. Its modules are (i) the coastal ecosystem, which embraces land, salt and fresh waters and the near atmosphere, and (ii) the human coastal community. As a result, its structure includes (i) the abiotic elements forming the niche of the ecosystem, (ii) the biotic elements of the ecosystem, and (iii) the social elements.

The coastal system is also a *complex system*. Complex systems differ from systems because they cannot be *described* exhaustively. They can only be *represented* in a holistic way (see Table 1.3), which implies using a model. The model is the interface between the general view of reality, which we have in our mind, and reality. This means that the individual model is a mediative element between a general model, designing the general system, and reality, consisting of individual systems.[7] In our case, the general model is sustainable coastal development, which requires the existence of local systems organised in such a way as to pursue the integrity of the ecosystem, economic efficiency and social equity. The local model is concerned with coastal systems pursuing sustainable development by practising ICM.

Managing the Coastal System: The Approach by FAO
The concept of coastal system adopted by FAO includes the following "principles and premises":
a) The coastal area is a unique resource system which requires special management and planning approaches;
b) Water is the major integrating force in coastal resource systems;
c) It is essential the land and sea uses be planned and managed in combination;
d) The edge of the sea is the focal point of coastal management programmes;
e) Coastal management boundaries should be issue-based and adaptive;
f) A major emphasis of coastal resource management is to conserve common property resources;
g) Prevention of damage from natural hazards and conservation of natural resources should be combined in ICZM (integrated coastal zone management) programmes;
h) All levels of government within a country must be involved in coastal management and planning;
i) The nature-synchronous approach to development is especially appropriate for the coast;
j) Special forms of economic and social benefit evaluation and public participation are used in coastal management programmes;
k) Conservation for sustainable use is a major goal of coastal resource management;
l) Multiple-use management is appropriate for most coastal resource systems;
m) Multiple-sector involvement is essential to sustainable use of coastal resources;
n) Traditional resource management should be respected;
o) The environmental impact assessment approach is essential to effective coastal management.
(From Clark J.R. [1992: 48–66]).

When a programme for ICM is undertaken, first of all the concepts of sustainable development and integrated management should be assumed as points of reference. Consequently, the general model of the sustainable coastal area is considered as the objective towards which the coastal system is expected to be driven. This model consists of an area where the integrity of the ecosystem is protected and the efficiency of the economy (optimisation of the renewable resources, increase in value of human resources and their cultural heritage, etc.) and social equity are pursued. In its turn, the coastal area is assessed bearing in mind the need to know all the elements required to set up management aimed at moving the area towards the goal. In this context:

a) the ecosystem is considered with the aim of understanding how and to what extent it can remain unaltered by the impacts from the local community and provides the useful natural resources for development;

b) the coastal economic and social organisation, based on the coastal use structure, is identified and investigated with the aim of understanding whether it directs the coastal system effectively towards sustainable development;

c) the role of the decision-making systems is taken into account to understand whether and to what extent they are able to effectively manage the coastal system. The ICM programme is designed on the basis of this assessment.

These concepts will be covered in more detail later under the following subject areas: coastal ecosystem (Chapter 3), coastal organisation (Chapter 5) and coastal system (Chapter 6).

1.7.   The Mediterranean, a Basin of Primary Interest

In order to complement the conceptual and methodological frameworks with the presentation of case studies, attention will be paid to the Mediterranean and its prospects of pursuing sustainable development (Chircop 1992). Two basic reasons justify this. Firstly, the coastal areas of the Mediterranean Sea are characterised by such strong human pressure and intense resource use that all the concerns arising when ICM is pursued are clear. Secondly, this was the first basin in the world where the coastal states have co-operated in protecting the environment. That approach, which has developed under the umbrella of UNEP, also includes the management of coastal areas.

A brief general presentation is useful to introduce the single aspects of the Mediterranean, which will be dealt with in the following Chapters. Only two aspects of close relevance to coastal management will be considered: (i) the Mediterranean diversity, which is the background to which each coastal management programme has to refer; (ii) the political and managerial mechanisms of the Mediterranean co-operation.

The Mediterranean Diversity
The Mediterranean region is characterised by unprecedented diversity: biological, social and cultural. As regards its marine waters, *biological diversity* is due to the coexistence of about five hundred living species. Over the last six million years most of them have migrated to these waters from the Atlantic and Indian oceans, when the

Mediterranean was open, and became acclimatised during those phases when it was left closed. At present organisms are converging from the Red Sea and Indian Ocean through the Suez Canal to the Mediterranean.

TABLE 1.5. The Mediterranean. Main data

Features	Unit	Data
Sea surface	million km^2/million sq m	2.5/0.965
East—west extent	miles/km	2,500/4,000
North—south extent	miles/km	500/800
Average water depth	km/yd	1.5/1,640
Water depth, Strait of Gibraltar	meters/feet	97/320
Water depth, Bosporus	meters/feet	70/200
Coastline total length	thousand km	45.0
Coastline island length	thousand km	17.7
MEDITERRANEAN coastal region, surface	million km^2/million sq m	1.5/0,579
Countries (*)	no.	21
Autonomous territory	no.	1 (**)
Urbanisation	% of the coastline	65
Coastal population, total Mediterranean, 1980	million inhabitants	84.5
Coastal population, total Mediterranean, 2000	million inhabitants	123.7
Coastal population, total Mediterranean, 1980/2000 increase	Percent	46
Population pressure, 2000 (residents and tourists)	thousands per km of coastline	5,700 to 6,600
Population pressure, 2025 (residents and tourists)	thousands per km of coastline	11,000 to 12,200
Power plants: existing/planned/total—Mediterranean	no.	112/43/155
Power plants, existing/planned/total—northern side	no.	60/4/64
Power plants, existing/planned/total—southern side	no.	52/39/91

* Albania, Algeria, Bosnia and Herzegovina, Croatia, Cyprus, Egypt, France, Greece, Israel, Italy, Lebanon, Libya, Malta, Monaco, Morocco, Slovenia, Spain, Syria, Tunisia, Turkey, Federation of Yugoslavia (Serbia and Montenegro).
** Gaza Cis-Jordan entity.
From MAP publications and Pavasovic (1996).

In addition, most of these numerous species are rare, and so, unlike the North Sea and other semi-enclosed European seas, the Mediterranean is endowed with only a few abundant species. These two factors, namely, the vast presence of exogenous species and the deficiency of abundant species, have converged to give shape to an unusual marine biodiversity. Moving landward from the shoreline numerous terrestrial ecosystems extend from the Mediterranean bush to the forest and are mostly characterised by a great variety of flora and fauna and high ecological fragility.

The *economic* and *social diversity* is undoubtedly based on historical roots, expressed by the vast number of cultural areas which were born and developed during history. Furthermore, this framework has become increasingly complicated because of two processes. On the one hand, the economic growth of the European Union (EU) has caused an increasing gap between developed and developing regions deeply influencing the inter-regional division of labour. A limited group of coastal regions endowed with advanced structures, such as technological core areas, hightech industries and logistic platforms, clash with numerous regions where the technological endowment is poor and the economic growth level is low. On the other hand, the economic gap between the EU and many other coastal and island regions has provoked intense migrations, especially from southern and eastern to the northwestern regions, highlighting the cultural diversity of the urban communities of Italy, France and Spain. The political struggle, primarily terrorism (1970s and 1980s) and conflicts in the former Yugoslavia (1990s), have accelerated migrations and increased the subsequent social changes.

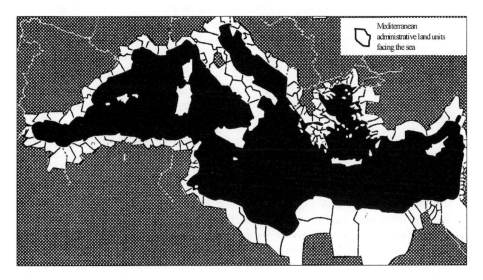

Figure 1.3. The Mediterranean coastal area, according to UNEP/MAP (1989, 187). It is considered as coinciding with the coastal administrative areas and the island endowment of the basin.

For *cultural diversity* it is sufficient to remember that the Mediterranean was the birthplace of the world's three monotheistic religions and that in the hands of its human communities there are the speculative scientific and ethical backgrounds which have deeply influenced the history of mankind. On this fertile ground, cultural diversity on the local scale has developed over time to the point that the Mediterranean coastal area and small islands now form an extraordinary and complicated web of *cultural units*. They are each imbued with its own history, traditions, land use patterns, social behaviour patterns and ways of perceiving the world and human existence. Globalisation, mainly as regards the standardisation of economic organisational patterns, settlements, and social behaviour, have profoundly jeopardised the cultural identity of these local communities. As a consequence, coastal management, as well as that of small islands, has the necessary task of preserving these identities as essen-

tial resources for sustainable development and as a basic heritage for future generations.

Co-operation in the UNEP Framework
As is well known, UNEP was established in 1972 as a result of the United Nations
Conference on the Human Environment, convened in the same year. Between 1972
and 1974 the Governing Council of UNEP "repeatedly endorsed a regional approach
to the control of marine pollution and the management of marine and coastal resources
and requested the development of regional action plans" (UNEP 1982: i). The Regional Seas Programme was initiated as the main outcome of that resolution and was
"conceived as an action-oriented programme having concern not only for the consequences but also for the causes of environmental degradation and encompassing a
comprehensive approach to combatting environmental problems through the management of marine and coastal areas" (*Ib.*). Moving from the Regional Seas Programme
(UNEP 1982), the Mediterranean Action Plan (MAP) was launched in 1975 at the
conclusion of an inter-governmental meeting convened in Barcelona by the Executive
Director of UNEP. Discussion focused on four main aspects:
a) Integrated planning of the development and management of the resources of the
    Mediterranean basin;
b) A co-ordinated programme for research, monitoring, exchange of information and
    assessment of the state of pollution and protection measures;
c) A framework convention and related protocols for the protection of the Mediterra-
    nean environment;
d) Institutional and financial implications of the Action Plan.
    As its conclusion, the meeting recommended developing programmes not only to
protect the environment but also to promote economic co-operation in those areas
which were closely concerned with environmental management.

The Barcelona Convention
In 1976 the Mediterranean States, convened in Barcelona, adopted the *Convention for
the Protection of the Mediterranean Sea against Pollution*. The Convention aimed at
pursuing pollution-referred goals, namely, a subset of goals designed by the 1975
Conference from which the Mediterranean Action Plan arose. It was followed by the
adoption of five protocols (Table 1.6).
    In 1993 the Eighth Ordinary Meeting of the Contracting Parties to the Barcelona
Convention (Antalya, Turkey) decided that the Mediterranean co-operation should
conform to UNCED principles and guidelines, with special reference to Agenda 21.
According to that resolution and the subsequent work on it, both the Ninth Ordinary
Meeting and the Conference of Plenipotentiaries to the Barcelona Convention were
convened in June 1995 with the aim of moving from the existing conventional approach to a new approach, consistent with Agenda 21 and other UNCED materials
and of pursuing sustainable development in the Mediterranean (Ruiz 1996). The Parties took two kinds of decision, consisting of adoptions and acquisitions.[8]

TABLE 1.6.   The Barcelona Convention system

Convention and protocols	Adoption and place	Status development (*)
Convention for the Protection of the Mediterranean Sea against Pollution Responsible: The MAP Co-ordinating Unit	Barcelona, Spain	a) 1976 b) 1978 c) 22
Dumping protocol Protocol for the Prevention of Pollution of the Mediterranean Sea by Dumping from Ships and Aircraft Responsible: The MAP Co-ordinating Unit, Athens	Barcelona, Spain	a) 1976 b) 1978 c) 22
Emergency protocol Protocol concerning Co-operation in Combating Pollution of the Mediterranean Sea by Oil and Other Harmful Substances in cases of Emergency Responsible: REMPEC, Malta	Barcelona, Spain	a) 1976 b) 1978 c) 22
Land-based protocol Protocol for the Protection of the Mediterranean Sea against Pollution from Land-Based Sources Responsible: MEDPOL, Athens	Athens, Greece	a) 1980 b) 1983 c) 22
SPA Protocol Protocol concerning Specially Protected Areas and Biodiversity in the Mediterranean (**) Responsible: Specially Protected Areas, Regional Activity Centre, Tunis	Geneva, Switzerland	a) 1982 b) 1986 c) 22
Offshore Protocol Protocol for the Protection of the Mediterranean Sea against Pollution resulting from Exploration and Exploitation of the Sea-Bed and its Subsoil Responsible: MEDPOL, Athens	Madrid, Spain	a) 1994 b) not yet c) 11
Convention on the Protection of the Marine Environment and the Coastal Region of the Mediterranean Sea Resulting from the amendment of the 1976 Convention	Barcelona, Spain	a) 1995 b) not yet c) 22
Hazardous waste protocol Protocol on the Prevention of Pollution of the Mediterranean Sea resulting from the Transboundary movements of Hazardous Wastes and their Disposal. Responsible: not yet decided.	Izmir, Turkey	a) 1996 b) not yet c) 11

*   *Codes*: a) approval, b) entering into force, (c) number of ratification (re: 1996).
** This protocol is resulting from the amendment of the Protocol on Specially Protected Areas by the Ninth Ordinary Meeting of the Barcelona Convention (1995).
Adapted from Pavasovic (1996) and http://www.unep.org/unep/program/natres/water/regseas.

TABLE 1.7. Evolution of MAP organisation

Year	MAP body	Place
1975	Adoption of MAP by 16 states	Barcelona, Spain
1976	Adoption of the Convention on the Protection of the Mediterranean Sea against Pollution by 16 states and European Economic Community (EEC).	Barcelona, Spain
1976	Launching of MEDPOL, Co-ordinated Mediterranean Pollution Monitoring and Research Programme	Barcelona, Spain
1976	Establishment of the Regional Oil Combating Centre (ROCC), today Regional Marine Pollution Emergency Response Centre (REMPEC)	La Valletta, Malta
1977	Establishment of the Blue Plan/Regional Activity Centre (BP/RAC)	Sophia Antipolis, France
1977	Establishment of the Priority Actions Programme/Regional Activity Centre (PAP/RAC)	Split, Croatia
1979	Establishment of the Mediterranean Trust Fund (MTF)	= =
1980	Establishment of the MAP Co-ordinating Unit	Geneva, Switzerland
1982	Removal of Co-ordinating Unit	from Geneva, to Athens, Greece
1987	Establishment of the Specially Protected Areas/Regional Activity Centre (SPA/RAC)	Tunis, Tunisia
1989	Establishment of the Secretariat of the Protection of Coastal Historic Settlements Centre (HS/C)	Marseilles, France
1993	Establishment of the Environment Remote Sensing/Regional Activity Centre (ERS/RAC)	Scanzano, Italy
1996	Establishment of the Mediterranean Commission on Sustainable Development (MCSD).	Athens, Greece

Adoptions. They adopted:
a) the Convention for the Protection of the Marine Environment and the Coastal Region of the Mediterranean, resulting from the amendment of the 1976 Convention;
b) amended protocols of the Barcelona convention (Table 1.6).

Acquisitions. The following documents were regarded as leading the Mediterranean co-operation based on the new Convention:
a) the Mediterranean Action Plan, Phase II (UNEP/MAP, 1995), presenting the initiatives expected to be carried out by the MAP Co-ordinating Unit from 1996 to 2005;
b) the Priority Fields, an extensive document illustrating the actions which will be regarded as of primary relevance to future policy;
c) the resolution on the establishment of the Mediterranean Commission on Sustainable Development;
d) the Agenda 21 for the Mediterranean (Agenda Med 21), which was agreed at the conclusion of an International (intergovernmental) conference in Tunis, November 1994 (Republic of Tunisia, Ministry of Environment and Land Use Planning, 1994).

The organisation of MAP consists of three components: (i) the Co-ordinating Unit (Athens, Greece), which depends on the UNEP headquarters (Nairobi, Kenya); (ii) the Programmes, consisting of sets of actions focusing on an individual core issue of the Mediterranean environmental protection; (iii) the Regional Activity Centres (RACs), consisting of branches of MAP dealing with tasks imposed by the provisions of the Convention and subsequent protocols.

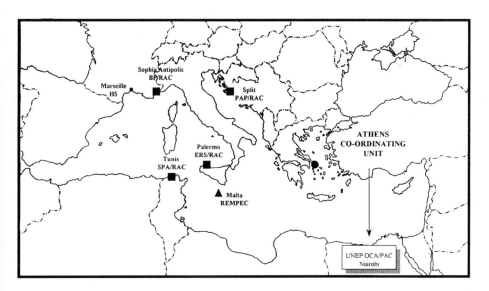

Figure 1.4. The organisation of the Mediterranean Action Plan (UNEP/MAP). *Acronyms*: BP, Blue Plan; ERS, Environmental Remote Sensing; HS, Historical Sites; OCA/PAC, Oceans and Coastal Area Priority Actions Centre; PAP, Priority Actions Programme; RAC, Regional Activity Centre; REMPEC, Regional Marine Pollution Emergency Response Centre; SPA, Specially Protected Areas.

1.8.   Coastal Management in the Mediterranean: A Stage-Based Model

The Convention and MAP system has provided the background and political justification for undertaking coastal area management programmes in the Mediterranean. This action was carried out by the Priority Actions Programme (PAP/RAC), a Regional Activity Centre (Split, Croatia), and has evolved through various phases. First, experimental activity was performed in the late 1980s to deal with issues on local environmental protection and coastal use development, and to test methods of assessing and designing management approaches. This activity was concerned with Croatia, Turkey, Syria and the Island of Rhodes and prepared the undertaking of coastal management programmes (Table 1.8).

1.8.1. PRE-TAKE OFF PHASE: 1989–1993

In 1989, the Sixth Ordinary Meeting of the Contracting Parties to the Barcelona Convention endorsed MAP to pilot *coastal area management programmes* (CAMPs) in the developing Mediterranean countries. In accordance with that resolution, PAP/RAC undertook CAMPs in those areas where the mentioned preparatory investigations had been carried out (Pavasovic 1992).

That exercise was very useful since it made it possible to design and test methods, and to test the reaction from local authorities and communities to initiatives aimed at protecting the environment and encouraging economic development.

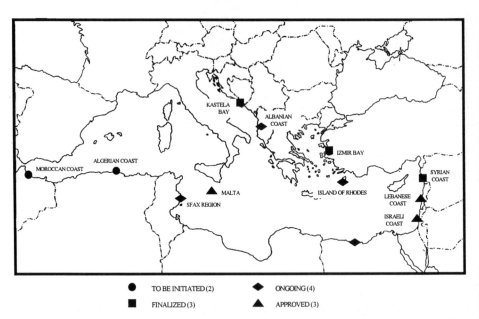

Figure 1.5. Coastal area management programmes of the Mediterranean Action Plan. Programmes. From UNEP(OCA)MED IG/Inf.3 (1995: 87).

1.8.2. TAKE OFF PHASE: 1993–1995

The Eighth Ordinary Meeting of the Barcelona Convention decided to undertake coastal management programmes in various parts of the Mashrek and Maghreb areas and in Albania. It was also decided to implement the management programme of the Island of Rhodes. These programmes were conceived as the first steps of the Mediterranean coastal management programmes tailored to Agenda 21.

These programmes were conceived as the first step of systematic Mediterranean co-operation on coastal management consistent with the guidelines from Agenda 21,

Chapter 17, and sensitive to the local environmental and social frameworks in southern coastal and island areas.

TABLE 1.8.   Mediterranean Action Plan. Coastal area management programmes

Experimental, preparatory activity and pilot projects	
1987–1988, pilot projects. Main tasks: assessment and management of individual resources, integrated planning studies, capacity building.	
Places: Kastela Bay (Croatia), Izmir Bay (Turkey), Island of Rhodes, Syrian Coast	
Pre-take off phase	
1989–1993, Coastal Area Management Programmes (CAMPs)—1st generation:	
The Syrian Coast, Syria	1989–1992
The Izmir Bay, Turkey	1989–1993
The Kastela Bay, Croatia	1989–1993
The Island of Rhodes, Greece, 1st phase	1990–1993
Take off phase	
1993–1995, Coastal Area Management Programmes (CAMPs)—2nd generation:	
The Albanian Coast, Albania	1993–1995
The Island of Rhodes, Greece, 2nd phase	1994–1996
The Region of Fuka, Egypt	1994 onwards
Sfax, Tunisia	1995 onwards
Drive to maturity	
1995 onwards, Coastal Area Management Programmes (CAMPs)—3rd generation:	
Preparation of programmes for Algeria, Israel, Lebanon, Malta, Morocco. Expected to start in 1997.	
Adapted from Pavasovic (1996).	

1.8.3. DRIVE TO MATURITY: 1995 ONWARDS

The experience gained in the early 1990s enabled UNEP to provide guidelines and recommendations on ICM with special reference to the Mediterranean (UNEP, 1995). That document was adopted in 1995 by the Ninth Meeting of the Contracting Parties to the Barcelona Convention and has been regarded as a main basis to move along three tracks: (i) to stimulate states and local decision-making centres to undertake ICAM programmes; (ii) to optimise the driving role of PAP/RAC and its capacity to provide technical assistance in the field; (iii) to conform methodological approaches and techniques to the UNEP guidelines and thereby to move towards the adoption of programmes and measures tailored to the principle of ICM (Vallega 1995 and 1996).

Currently, a wide network of initiatives is being developed to carry out ICM programmes. These are concentrated in the southern and eastern sides of the basin, where they are supported by inter-governmental organisations, especially by the World Bank. Co-ordination assigned to PAP/RAC, as well as other Programmes and Regional Activity Centres (RACs) of MAP are expected to be deeply involved. This is the case of the MEDPOL Programme, Phase III, for the protection of the sea from land-based activities, and the Specially Protected Areas (SPA, Tunis), a RAC designed to pursue the conservation of one hundred fragile local ecosystems along the Mediterranean coasts.

1.9.   Basic Questions and Key Issues

The approach to ICM, accomplished in this Chapter, has demonstrated that ICM is a very intriguing subject. The most diverting path should be to regard it as a merely technical task. Such a perception would have been possible during the take off and drive to maturity phases of coastal management (Section 2, Table 1.2), but not now as coastal management programmes need to be designed bearing in mind global change and to be tailored to the sustainable development principle. As a result, before moving along this path, the awareness that conceptual and methodological aspects are to be taken into account is a precondition of success. It is particularly important to recognise that the holism principle, recommended by Agenda 21 and other institutional materials, requires the adoption of a constructivist logical background (Table 1.3). On this basis, the coastal planner and manager could usefully drive their work and effort considering this decalogue:

1. designing a holistic view of the coastal area and its organisation (the coastal system) by providing a stage-based view of its evolution;
2. assessing what interaction the coastal areas has established with its external, physical and social, environments;
3. focusing on the expected impacts from climate change;
4. in this framework, designing the web of relevant core issues and main prospects;
5. referring prospects to the concept of sustainable development using rigorous criteria and critical views;
6. considering ICM as the tool to drive the coastal system towards sustainable development and designing the subsequent possible spectrum of goals;
7. exploring pertinent scientific and institutional literature to draw orientations and guidelines respectively;
8. drawing models of the social perception of the coastal area needs;
9. drawing lessons from the experiences made in other coastal areas;
10. using proper technical tools to assess the present coastal organisation and to design possible future organisational patterns.

CHAPTER 2

THE COASTAL AREA IN GLOBAL CHANGE

The approach presented in the previous Chapter was concerned with the sustainable development principle. The concept of global change, closely linked with that of sustainable development, will be now introduced and the following questions will be dealt with:

Questions	Figures	Tables
How should global change be conceptualised?	2.1, 2.2	
Is a stage-based view useful to design the economic and social organisation of the coastal area?	2.3	
Which economic and social processes should be focused on by coastal managers and planners?		2.2, 2.3
How climate change and sea-level rise could be assumed as a basis for coastal management?		2.2, 2.3
What methodological approach could usefully be adopted to deal with global change and related biogeochemical processes?		2.4

2.1. The Global Change Concept

The sustainable development principle is based on a double awareness: any approach to management of the environmental and social contexts has to move from the concept of global change; development and social equity are to be put at the heart of any sustainable development-aimed programme. The primary factor of global change is the warming of the near atmosphere and subsequent climate change. Because of the so-called greenhouse effect[9], warming has been accelerating especially since the mid 20th century. Coastal areas have been influenced more than many other parts of the world by a chain of relationships, which has produced cumulative effects:

- the economic and social organisation, born as a consequence of the First Industrial Revolution, has speeded climate change;
- climate change has brought about sea-level rise in many latitudes, but particularly in the inter-tropical where most of the less developed countries are located;
- this has intensified the economic and social change, and
- has caused stress in many parts of the world, particularly within the inter-tropical zone.

As a result, the need to mitigate the human-induced acceleration of climate change is intimately associated with that to successfully tackle economic and social development. These aspects relate to coastal management because the coastal areas form the

environment most profoundly influenced by global change and are subject to an unprecedented increase in human pressure.

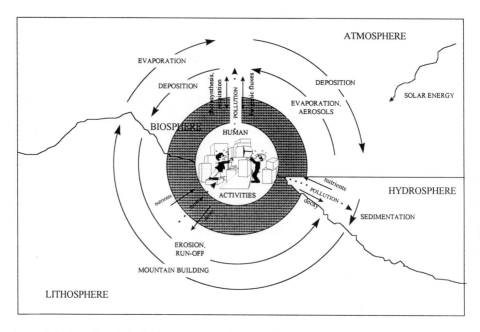

Figure 2.1. The view of the Earth system, according to the International Geosphere-Biosphere Programme
(IGBP). Adapted from ICSU (1996; 9).

"'Climate change'"—the Convention on Climate Change (1992, Article 1.2)
states—"means a change of climate which is attributed directly or indirectly to human
activity that alters the composition of the global atmosphere and which is an addition
to natural climate variability observed over comparable time periods". This motivates
considering climate change as the basis to re-orientate and restructure the main patterns of natural resource use and the criteria to manage them. The relevant approach
has been developed by the *International Geosphere-Biosphere Programme* (IGBP), the
most important research programme launched by UNESCO (1986).[10] Nevertheless, as
climate change is being accelerated by humans, the concerned economic and social
processes are to be taken into account. These aspects are basically dealt with by the
Human Dimensions of Environmental Global Change Programme (HDEGCP, called
more briefly HDP), another UNESCO programme launched in 1988 to complement
IGBP. As a final result, the assessment accumulated since the mid-1980s has focused
on the feedback between human communities and the ecosystem and has provided useful frameworks for undertaking sustainable development-consistent actions tailored to
environmental and resource use management (Carter 1987; Carey and Mieremt 1992;
Jelgersma and Tooley 1992; Peck and Williams 1992; Pernetta and Elder 1992;
Hoozemans, Stive and Bijlsma 1993; Vellinga and Klein 1993; Nicholls, Leatherman,

Dennis and Volonté C.R. 1995). As regards climate change, research is concerned with the period since the last small Ice Age, that is the past 18,000 years. As regards the generation of human influences, as the initial effects of the industrial revolution are the starting point, the processes developed during the last 150 years are the prime reference.

The Basic Approach to Global Change
"Human activities have become a significant force affecting the functioning of the Earth system. Our use of land, water, minerals and other natural resources has increased more than ten-fold during the past 200 years, and future increases in population and economic growth will intensify such pressure. Large scale (and often novel) chemical transformations are involved, together with changes to the natural transfers of energy and materials around the world.

Impacts on the dynamics of the biosphere are of particular concern. Climate, the global cycles of carbon and water, and the structure and function of natural ecosystems are all closely linked. Major changes in any one of these systems will affect the others, with potentially serious consequences for humanity and other life on Earth".
(From IGBP [1990], *Reducing Uncertainties*. Stockholm, IGBP, Bulletin no. 4).

Physical Climate System

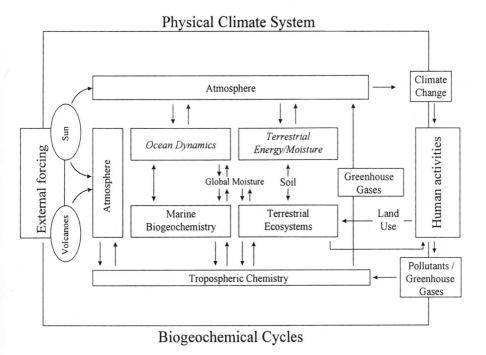

Biogeochemical Cycles

Figure 2.2. The model of global change adopted by the International Geosphere-Biosphere Programme (IGBP). Adapted from ICSU (1993: 32).

Change is global because it has not only involved the near atmosphere, but also all the biogeochemical cycles, influencing the economic and social organisation which has been the main cause of accelerating climate change. As a result, since the First

Industrial Revolution a circular relation has been materialising involving every scale: (i) the global scale when the Earth' ecosystem is considered as a whole; (ii) the regional scale where attention is referred to a wide, frequently multi-national, geographical space, such as the Mediterranean; (iii) the national, and (iv) subnational (also called "local") scales.

The global change concept attributes great importance to variability, considering it a factor making it hard to design the evolution of the ecosystems: "Key interactions in the Earth system, especially biological responses to physical and chemical changes in the environment, are typically non-linear. This becomes particularly important when we consider that the anticipated global warming of the next half century could be of the same magnitude as that experienced by the Earth's ecosystems since the last glacial maximum 18,000 years ago. The species of animals and plants that have survived in temperate and subtropical latitudes over the past few million years have successfully adapted in numbers and geographic range as climate has warmed and cooled. The climate conditions projected for the near future are beyond the range of experience of these organisms, and the rates of change involved may be unprecedented" (IGBP 1988, 4: 15). This approach is invaluable for coastal management since it leads managers and planners to avoid mechanistic and deterministic approaches to the ecosystem. To consider the future of the physical environment as evolving non-linearly, and the evolution of the ecosystem as unpredictable, is a very intriguing exercise, not least because it requires setting up management programmes and plans supposing a future that cannot consist of the projections of present trends. However, this exercise is unavoidable: its practice marks a threshold between the conventional and innovative approach to the coastal systems.

To move from theoretical to empirical approaches, and to present how global change might be assessed in the context of ICM programmes, the Mediterranean will be considered in the following sections. Attention will critically focus on economic and social change, and then shift to the change in climate and ecosystem. *The resulting approach may be regarded as a tentative model of "Global social and natural change of the coastal area" to be applied in the initial step of the ICM programme construction.*

2.2.   Mediterranean Human Change

2.2.1. FROM THE NEO- TO THE TRANS-INDUSTRIAL STAGE

Approaching the twenty-first century we are aware that a historical bifurcation took place in the mid-1970s consisting of the transition to a new stage. The previous one, which was collapsing during that decade, may be called "neo-industrial", since it developed during the twentieth century as the modern evolution of the palaeo-industrial stage, generated by the First Industrial Revolution. The present stage, the main features of which have become self-evident in the 1990s, may be defined "trans-industrial" since its economic and social organisation has caused the separation between the industrial and tertiary sectors to collapse, and has therefore transcended the past economic and social organisation. This is the main reason why some authors have regarded the mid-1970s as the watershed between the modern and post-modern societies (Harvey 1990). The trans-industrial stage was generated by profound changes in the *social division of labour*, mainly due to the progress in computer science and techniques and to the role of new disciplines, such as bio-engineering. These processes have been associated with changes in the *international division of labour*, mainly due to the consequences of de-colonisation and the arousal of the environmental debate. The take off of the new stage lasted about twenty years. Its maturity phase began in the mid-1990s primarily because of the revolutionary consequences of globalisation processes and the adoption of sustainable development by the international community. As can be seen, the need to set up coastal management programmes took shape during the take off phase of the trans-industrial stage and the prospect of building ICM programmes has come to the fore during its maturity (Figure 2.3).

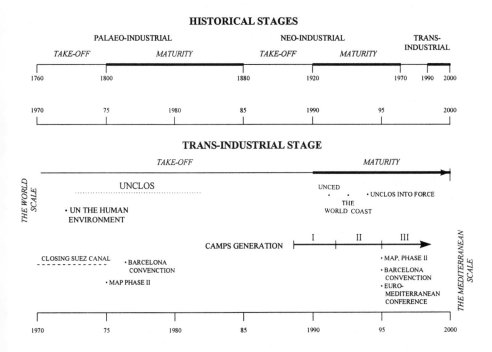

Figure 2.3. The stage-based model of economic and social change. *Acronyms*: CAMP, Coastal Area Management Programme (of MAP); MAP, Mediterranean Action Plan; UNCLOS, United Nations Convention of the Law of the Sea; UNCED, United Nations Conference on Environment and Development.

As for the Mediterranean, 1975 could be regarded as a very significant moment for the take off of the new stage.[11] It was characterised by two events, both influencing multi-national co-operation and coastal management. *First*, the Suez Canal, which had been closed in 1967 on account of the third Israeli-Arab war, was reopened. The routes linking the Mediterranean with the Red Sea, Indian Ocean, south-eastern Asia and Far East were re-established, the historical eastbound gateway from the Black Sea was restored and economic interests, presupposing the use of the Suez waterway, acquired new importance to the point of acting as a main tool to prevent strong conflicts on the Mediterranean side of the Near East. *Secondly,* the adoption of the Mediterranean Action Plan (1975), followed by that of the *Convention for the Protection of the Mediterranean against Pollution* (1976), strengthened the co-operation prospects even though the Mediterranean was troubled by geopolitical stress, including terrorism.

2.2.2. TAKE OFF OF THE TRANS-INDUSTRIAL STAGE

During the take off of the Mediterranean trans-industrial stage (1975–1995) many changes took place. As regards those influencing the coastal regions and islands, it is worth considering the geopolitical framework, which was characterised by some very influential processes: (i) in 1981 the European Economic Community (EEC) enlarged southwards[12]; (ii) as a consequence of the collapse of the Soviet Union, the Mediterranean was no longer an area of east-west political and military confrontation; and (iii) ethnic conflicts, together with conflicts within the Arab world, replaced the large, ideologically-sustained conventional conflicts, while the Israeli-Palestinian confrontation entered a peace-oriented phase despite the continuous stress.

The subsequent profound economic and social changes have generated long-term impacts on coastal areas. *Inter alia,* the following are worth emphasising.

Urban Pressure
During the 1970s and 1980s coastal urban settlements mostly expanded in the northern Mediterranean side, especially in the EU area. Small megalopolises took shape, particularly in the northwestern Mediterranean, from the North Tyrrhenian Sea to the sea facing the Cataluña. Since the mid-1970s, urban growth has been accelerating in many areas of the African and Asian sides of the basin, while coastal urbanisation has been decelerating in the north. In the meantime migrations from the southern side to the EU side have expanded.

Coastal Industrialisation
During the neo-industrial stage many manufacturing coastal areas were created within or near the seaports of Italy, France and Spain to process energy sources and raw materials and to convey the resulting products to inland regions. A large part of raw materials, mainly hydrocarbons and fertilisers, were imported from the southern Mediterranean. As a consequence, the Mediterranean seaport endowment was marked by the development of loading bulk terminals in the South and unloading terminals in the North. During the late 1970s and 1980s that pattern of the Mediterranean division of labour began to change because, in the northern side, iron and steel industries, jointly with oil refineries and petrochemical plants, declined and some industrial areas

collapsed. A new inter-regional division of labour materialised consisting in the re-location of raw material processing plants from North to South. This process gener-ated industrial growth in many parts of the African (Tunisia, Algeria and Libya) and Asian (mainly Turkey and Egypt) regions. The newly-built plants evolved over time conveying semi-finished products to both domestic and foreign markets. During the 1970s, offshore oil and gas exploitation started and spread throughout the Adriatic and Ionian Seas, the Balearic Sea and Tunisian, and Egyptian waters. This process led to the perception that the Mediterranean would acquire an important role in that new in-dustrial field.

Tourism and Recreational Uses
The development of tourism and recreation may be regarded as a leading process, having had a greater influence on the Mediterranean coastal organisation than that from changes in industrial areas. The development of tourism on the African and Asian sides, the unexpected involvement of islands in this process, and the widespread rise in cruise-based recreational activities were the most significant manifestations. In the 1980s the need to develop environmentally- and culturally-sound tourism was re-garded as important by an increasing number of decision-making centres. That per-ception has led to the creation of coastal marine parks and reserves, the adoption of measures to protect endangered species and the growing archaeological interest from local and international tourists.

2.2.3. THE MATURITY OF THE MEDITERRANEAN TRANS-INDUSTRIAL STAGE

In the 1990s the trans-industrial stage entered its maturity phase, and once again the major signal was political. In 1995 the adoption of the *Convention for the Protection of the Marine Environment and the Coastal Region of the Mediterranean*, legally claimed as an amendment of the 1976 Convention and associated to the Mediterranean Action Plan, Phase II, inaugurated the Mediterranean co-operation aimed at pursuing sustainable development. During this time the conflicts involving the divided Yugo-slavian state, which started in relation to the end of the Cold War, ceased and a wide co-operation programme between EU and the extra-EU areas of the Mediterranean started up (*Euromediterranean Conference*, 1975). The following year the *Mediterra-nean Commission on Sustainable Development* (MCSD) was established in the frame-work of the Barcelona Convention.

The political change triggered important social changes including two processes which have exerted great influence on coastal areas. *First*, the collapse of the Soviet Union and the fading of Communism opened the way to claims for independence by ethnic groups. Hence there arose the excitement of historical roots and cultural heri-tage and the design to preserve and exhibit them, as well as protect minority groups. Many of these processes are concentrated in coastal regions and islands because of the sea-oriented development of the Mediterranean indigenous communities. *Secondly*, the solemn principles regarding the protection of women and indigenous peoples, proposed in many parts of the UNCED materials, encouraged processes aimed at strengthening the role of women in communities, such as in Arab countries. As a result, social and ethnical groups have strengthened their role in the social organisation and change. These processes have special relevance to coastal areas and islands, basically on the African and Asian sides, where most domestic populations are located in coastal settlements.

Furthermore, economic change has been profound and closely linked with that of the geopolitical and social changes. At least four processes are worthy of consideration because of their capability to influence coastal organisation.

Coastal Industrialisation
During the maturity phase of the trans-industrial stage manufacturing development has progressed more in the South than in the North. More importantly, the growth of manufacturing plants in these regions, especially in Turkey and Egypt, has been characterised by the increasing production of finished goods. As a consequence, southern industrialisation has become more flexible and articulated associating conventional plants, processing and exporting local energy sources and raw materials, with plants also producing finished goods for both domestic and foreign markets.

Seaports
Industrial areas on the EU side have declined while some break-bulk seaport terminals have been abandoned. This decline has been associated with the increase in containerised traffic. New terminals and new seaports (such as Genoa Voltri, in the North-western Mediterranean) were built quite far from the historical settlements. These spatial processes, involving both industry and seaports, have made vast areas available for other facilities. The city, on the one hand, and industry and seaport on the other, have been liable to spatial dissociation which has somewhere also led to functional dissociation. These three components of the coastal system, which were integrated in the past to the point of acting as the engine of coastal development, have moved towards new organisational patterns whose final features are difficult to envisage.

Mineral Resource Use
This sector has undergone no profound changes. In particular, oil and gas exploitation has not grown and expanded as was expected in the 1970s. During the 1980s this production was limited to five countries (Egypt, Italy, Libya, Spain and Tunisia) and was less limited than in the past decades. Also proven reserves were not remarkable. During the first half of the 1990s no factors accelerating exploitation and production have arisen.

Fisheries
This use has undergone serious concerns because of two concurrent factors: the ecological degradation cf the marine biotic communities has persisted largely due to the persistent use of non environmentally-sound catch techniques; the saturation of biomass exploitation has caused an imbalance between the demand for fish and the pro-

duction capability of the marine ecosystem. Hence there are two main consequences: the progressive change in fishing techniques and the development and diffusion of aquaculture.

Recreational Uses and Tourism

These have not only benefited from uninterrupted growth in coastal and island areas, but have also become increasingly sea-oriented due to a wide range of factors. *Inter alia,* the increased proclivity to create marine parks has stimulated undersea exploration by specialised vessels and plants. The underwater archaeological remains, an increasing number of which had been discovered and used for tourism, have been perceived as a new frontier of sea exploitation. Yachting and cruise have surpassed all expectations.

Impacts on Coastal Management

It is now worth considering what prospects and concerns should be dealt with by the Mediterranean coastal managers and planners. Bearing in mind the mentioned geopolitical, social and economic processes, it is justified to state that sustainable coastal development may be achieved effectively where a wide spectrum of economic processes is put in motion including:

a) the design and operation of new patterns of *integration between industrial development and urban growth* based on multiple manufacturing functions and development of new high-tech tertiary sectors;
b) the diffusion of *maritime waterfront revitalisation* keen to use abandoned spatial and environmental resources and, contextually, to protect the cultural heritage (archaeological remains, historical centres, old port structures and others);
c) the diffusion of exploitation of *renewable energy uses*, such as those based on solar and wind sources:
d) the sustainable development of Mediterranean *horticulture and fruit-growing*;
e) the development of *aquaculture*;
f) the development of environmentally-sound *recreational coastal and marine uses*.

The more the prospects of sustainable development of the individual coastal areas are perceived as important by the local community, the easier it is to set up integrated management programmes. Prospects and relevant needs differ moving from developed to developing coastal regions and islands. In the northern Mediterranean, coastal management programmes are required:

a) to design adequate measures to convert abandoned manufacturing areas;
b) to provide settlements and services to immigrants from southern and eastern extra-EU regions;
c) to integrate, much more than in the South, resources from fisheries with resources from aquaculture;
d) to tackle the impoverishment of freshwater due to increased uses, peaks of demand for water during summer, water pollution and intrusion of saltwater into aquifers;
e) to restore destroyed and damaged coastal and marine ecosystems;
f) to protect the cultural heritage.

In the southern Mediterranean coastal management should focus on quite a different spectrum of needs including:

a) the conservation of historical settlements and other historical man-made structures;
b) the protection of cultural heritage against mass tourism and profit-oriented uses;
c) the optimisation of freshwater uses, especially in arid areas;
d) the preservation of marine and land ecosystems, especially the Mediterranean bush, dunes and desert;
e) the protection of local land and marine trophic webs including rare species.

The major issue involves the growing human pressure in many Mediterranean coastal areas. The rise in Mediterranean coastal population from 1980 to 2000 has been estimated at 46 percent (84.5 to 123.7 million inhabitants). An increase has characterised both northern and southern sides, although smaller in the North (19,7 percent; 51.6 to 61.8 million inhabitants) than in the South (88,2 percent; 32.8 to 61.9 million inhabitants). According to the Blue Plan[13], the coastal population in 2025 is estimated to be about 220 million inhabitants and the divergent trends of northern and southern sides are expected to continue. Population growth is associated with urbanisation: at the present time, 87 million coastal inhabitants live in cities and they are expected to double by 2025. National and international tourism is reckoned to reach 140–180 million people in 2000, and 220–340 million people in 2025 (Table 1.5). As a final result, the average population pressure, including residents and tourists, is expected to be 5,700 to 6,600 people per km of coastline at the end of this century, and 11,000 to 12,000 in 2025. It should double during the next generation, and mostly involve the southern side of the basin, islands and archipelagoes, and the eastern side of the Adriatic provided that peaceful conditions persist in former Yugoslavia.

2.3.   Mediterranean Climate Change

The presentation of geopolitical, economic and social change highlights how much human pressure will influence the evolution of the coastal ecosystem. As a consequence, managers and planners are inclined to wonder what impacts natural and human causes will have in accelerating global change throughout the Mediterranean. Attention therefore shifts back to climatic change, basically triggered by atmospheric warming.[14]

Since the end of the nineteenth century the temperature of the global near atmosphere has increased by about 0.5 °C and, according to the General Circulation Model (GCM), it is expected to increase further from 0.5°C to 1.4°C between 1990 and 2030. In the long term, Palutikof and Wigley (1996: 36-7) estimate that, during the twenty-first century, "temperature changes will be at least over the Mediterranean Sea and the adjoining coastlines of Italy, Greece and North Africa (...)". The scenario "would yield an actual warming range in 2100 of between a minimum of 1.5–3.4°C (...) increasing to a maximum of 2.2–4.9°C in the north-east".

Trends in rainfall, which are mostly influenced by warming of the near atmosphere, indicate that the Mediterranean as a whole should undergo a slight increase in precipitation. "Over the northern Mediterranean the mean changes are around 2–3%/°C global temperature change, rising to over 3%/°C in some areas. The only area of reduced precipitation lies over southern Italy and the adjacent parts of North Africa" (Palutikof, Wigley 1996: 39). The scenarios of temperature and precipitation indicate remarkably different trends from region to region. As a result, coastal management programmes have to move from the trend attributed to the whole Mediterranean to regional scenarios concerned with the influence from climate change to the individual areas.

Atmospheric Warming in the Mediterranean
"[Warming] has been neither continuous nor spatially homogeneous; trends varied substantially from region to region. Over the past 40 years, for example, the whole Mediterranean basin, along with much of the North Atlantic and western Europe, has undergone a noticeable cooling. That such a cooling has occurred in spite of the expected global-mean greenhouse-gas-induced warming demonstrates the degree to which external forcing influences can be masked by natural climate variability. This is especially so at the regional scale. Similar decade and longer time-scale "anomalies" (...) can be expected to occur in the future."
(From Wigley [1992: 17]).

Predictions of sea-level variations are methodologically difficult to make and results are less firm that those concerning temperature variations in the near atmosphere. This is mainly due to the complicated web of extra-atmospheric factors influencing the marine body, including tectonic changes, tidal variations, thermal expansion of sea waters, man-caused subsidence, and others. As regards the global scale, the Intergovernmental Panel on Climate Change (IPCC) estimated a mean sea-level rise of 31–110 cm by the end of the twenty-first century. This estimate was assumed as a basis for the Report to the World Coast Conference (Intergovernmental Panel on Climate Change 1993). As regards the regional scale, experts converge estimating that during the twentieth century Mediterranean waters have risen by 1.0–1.2 mm/yr and globally by around 12 cm (assuming a 0.5°C rise in temperature). The expected rise in the next three decades is estimated between 12 to 18 cm, except for areas—such Nile and Po deltas—prone to subsidence, where the rise could be as much as 25–40 cm (Milliman 1992). A rise of around 20 cm in the three next decades (re: 2020–2030) "would not, in itself, have a significant impact in the Mediterranean, except locally. However local sea-level changes could be substantially greater than this in some areas because of natural land subsidence, that may be enhanced by excessive groundwater extraction. The negative impacts will be felt in low lying areas, deltas and coastal cities" (Jeftic, Keckes and Pernetta 1996: 23). A wide range of impacts from sea-level rise is predicted to include: (i) the increased wave impact on exposed coasts, such as the Venice lagoon and Po delta; (ii) impacts on seaward prolonged port installations, such as in Alexandria; (iii) an increasing frequency and intensity of flooding of estuaries, canals and lagoons; (iv) possible impacts on coastal settlements, including historical remains, located near the coastline and subject to tidal variations; (v) accelerated coastal erosion and inundation stimulated by the increasing frequency of storms; (vi) accelerated sea water intrusion into coastal aquifers (*Ib.*: 23–5).

2.4.   Impacts on Brackish Resources and Land Use

A lack exists between precipitation intensity and frequency during the year, and freshwater resource use. Consequently, coastal managers are invited to consider as a model the feedback between climate change and the hydrological system, and the subsequent input to freshwater resource use. Many methods for using such a basic model for investigations on regional and local scales have been designed (Lindh 1992). Regarding coastal management three main consequences have acquired increasing importance: (i) saltwater intrusion into coastal areas; (ii) inundation and runoff in urban areas; (iii) impacts on water resource availability.

Saltwater Intrusion in Coastal Areas
As regards saltwater intrusion, it is necessary to consider estuaries and aquifers. Both components of the environment are strongly influenced by human-induced factors: *landwards*, the development of freshwater uses, provoking the weakening of aquifers and the subsequent intrusion of saltwater, and the exploitation of oil and gas fields, provoking subsidence; *seawards*, a sea-level rise. "Salt-water intrusion—Lindh notes (Lindh 1992: 59)—is not a new problem, as witnessed by the number of "salt-water intrusion meetings" held since 1968 (...). Today the consequences of the overexploitation of ground water from coastal wells are clearly understood. An interesting parallel to extreme extraction of water from wells is the effect of salinization in arid and semi-arid irrigation projects by ground-water use. In both examples, it has proved to be extremely difficult to renovate such salt-contaminated wells". Saltwater intrusion is a historical concern of many Mediterranean coastal areas but its importance has been increasing remarkably since the 1970s due to the associated impacts from the growing human pressure and the increase of the per-capita demand for water basically due to changes in social behaviour. According to Margat (1992: 36), in the whole Mediterranean coastal and island regions a great gap has opened between the northern and southern sides in terms of water exploitation ratio, intended as the ratio between (i) the ground water drawn off, and (ii) the potential of ground water. In the southern side the ratio rises to 74.5 while in the northern side it decreases 13.5. As a result, it should be expected that, as sea level rises and coastal human pressure increases, intrusion of saltwater in the aquifers will spread and expand up to ten kilometres from the coastline into the river water columns.

Inundation and Runoff in Urban Areas
Sea-level rise is expected to cause more frequent and more destructive inundation. Coastal lagoons, salt marshes, wetlands and brackish lakes will be subject to the subsequent effects. In the meantime many coastal and island cities will undergo the associated influences from the increase in precipitation due to climate change and sea-level rise. An increased runoff proportional to that of precipitation is expected. As a result, coastal managers and planners should focus their attention on the scenarios of these processes.

TABLE 2.1.   Brackish areas in the Mediterranean

Types of interface	surface km^2		
	northern side	southern side	total
Deltas:			
Nile		24.0	24.0
Rhone	2.1		2.1
Po	1.4		1.4
Ebro	0.5		0.5
Lagoons	2.9	2.7	5.6
Delta marshes	0.6	–	0.6
Total lagoons and deltas	3.5	2.7	6.2
Data from Margat (1992: 8).			

Impacts on Coastal Resources

No consensus in literature estimates freshwater use per capita. However, this is not a big concern in developing coastal management programmes since they refer to areas in which *ad hoc* investigations need to be carried out to acquire useful data. Bearing in mind the final objective of the ICM programme, i.e. sustainable coastal development, a wide range of freshwater uses have to be considered including: water treatment and domestic drinking supply; water supply for industry, agriculture and energy production; sewage, sewage treatment, effluent disposal and dilution; land drainage and food protection; navigation; fisheries, conservation and irrigation (Lindh 1992: 89). In the past, the demand per capita has increased more on the northern side, especially in the EU coastal regions, than on the southern one. Following the scenarios on climate change, the Mediterranean is expected to undergo less profound impacts than other regional seas yet, a new stage is taking shape characterised by higher freshwater use rates in the South than in the North.

In conclusion, the impacts from climate change on the water cycle and the subsequent effects on water resource uses require coastal management programmes in the Mediterranean to be sensitive to the possible local consequences, with special reference to the lowlands. Consequences are qualitative and quantitative: the former consisting of deterioring freshwater quality due to saltwater intrusion into groundwater, insalination and degeneration, the latter consisting of increased demand for freshwater due to the use growth.

2.5.   Impacts on Soils

Sea-level rise and coastal erosion will enhance soil vulnerability and accelerate soil degradation. With regard to the relevant web of physical and chemical processes, Imeson and Emmer (1992) suggest focusing on the salt balance of the soil, supply and mineralisation of organic matter and land degradation.

The Salt Balance of the Soil

Salt balance varies according to the frequency and intensity of precipitation, groundwater levels and other circumstances. A general decrease in precipitation or increase in evapotranspiration usually provokes an expansion of soils affected by saline or sodic conditions. Large areas characterised by annual rainfall less than about 600 mm could be seriously damaged by a decrease in precipitation because of the subsequent changes of physical conditions, including the exchangeable sodium percentage (ESP). These consequences have particularly affected many coastal areas of Spain and Italy. Small changes in ESP, for example due to a shift of its value at the end of dry season, could deeply influence soil sensitivity and have adverse effects on soil use. An increase in precipitation or decrease in evapotranspiration "could have a positive effect in areas suffering from sodic or saline conditions in cases where higher rainfall leads to better and deeper drainage" (Imeson and Emmer 1992: 104). However, this does not occur everywhere, and the possible impact on salt balance from climate change needs to be designed paying attention to the local conditions of each coastal area, namely, water balance, soil types, exposition and slope. In particular, salinity problems due to climate change are expected to become severe "in areas receiving between 300 and 600 mm of rainfall and would become less important in areas where precipitation dropped below this level. Such areas with generally irrigated agriculture of the occurrence of nitric and sodic horizons in the soil profile are not extensive in European and most other Mediterranean regions today but there is a large area potentially vulnerable to salt problems" (Ib.: 105).

Caliche Development

The variations in the crust of sodium carbonate, enhanced by climate change and subsequent variations in precipitation, are expected to influence the soil conditions of areas where rainfall varies from 100 to 500 mm and to be associated with desertification.

Supply and Mineralisation of Organic Matter

These are also a key issue because they are subject to impact from climate change and may cause impact on soil structure, so influencing agricultural productivity. At present, it is difficult to provide a global view of the Mediterranean and to estimate the impact expected from the near atmosphere temperature variations in the next decades. The main reason is that the local climatic conditions of the Mediterranean vary too much from region to region.

Fires

In many regions of the Mediterranean, especially in those mostly affected by increasing human pressure, the number of *fires* has increased in the last few decades. Higher temperatures, especially associated with dry conditions, will enhance this process causing two impacts: (i) the degradation of vegetation, especially the Mediterranean bush, (ii) the impoverishment of agriculture, due to the associated loss of organic matter and increase in runoff during rainfall in dry season.

Desertification

As far as soil sensitivity is concerned, "*desertification* means the diminution of biological potential and primary productivity due to climate change and excessive exploitation by man. Land degradation in arid and semi-arid Mediterranean regions (rainfall < 600 mm per annum) has strong analogies with (...). The sensitivity of desertification will be significantly increased by any increase of aridity, and this might accelerate or reinforce present or past degradation process" (Imeson and Emmer 1992: 113).

Land Degradation

This problem, which has always marked the Mediterranean, has two aspects: on the one hand, it is the heritage of the Mediterranean history, *inter alia* showing the evolution of human-ecosystem interaction; on the other hand, it is going to be a main consequence of climate change and subsequent variations in precipitation. Focusing on coastal management, a wide range of aspects are to be put on the agenda of managers and planners including: (i) *rainfall and runoff erosivity*, namely the ability of rainfall and runoff to generate or accelerate erosion; (ii) *erodibility of soils*, intended as the erodibility of materials subject to erosion, which is relevant where climate change and variations in rainfall bring about lowering of organic matter, mainly due to changes in concentration and composition of water-soluble salt; (iii) *soil erosion*, consisting of sediment yields from surface wash, which is estimated to be highest in areas receiving between 200 to 600 mm of rainfall per year; (iv) *sediment budget*, due to the associated consequences from climate change-caused erosion and measures undertaken by local communities to convey sediments.

As a conclusion, Imeson and Emmer (1992: 125) present this *breakdown of impacts* associated with an increase in the temperature of 3 to 4°C by the end of the twenty-first century:

a) an increase of aridity;

b) an increase in soil degradation due to the effects of increased temperature and changes in organic matter;

c) an increase in soil erodibility mainly due to lower rates of water acceptance;

d) higher rates of erosion on slopes and even more overland flow during extreme rainfall events;

e) higher sediment concentrations in slope runoff;

f) higher sediment and solute supply from badlands and/or marl areas;

g) an increase in the risk of rill and gully erosion;

h) higher sediment and lower runoff and sediment transport in drainage basins;

i) lower rates of sediment transport in the major rivers;

j) adjustment of river channels;

k) a higher susceptibility to forest fire and a decrease of biodiversity;

l) more arid soil conditions and a decrease in the areas suitable for profitable agriculture;

m) increasing management costs in agriculture to maintain crop yields.

2.6. Basic Questions and Key Issues

The *Report* to the *World Coast Conference* (Intergovernmental Panel on Climate Change 1993) emphasises a spectrum of questions which policy makers find themselves focusing on, such as "is it safe to build here, or will erosion increase?", and "is it safe to swim here, or is the water too polluted?". What is required is a long-term view of the climate variations and impacts on the abiotic components of the local ecosystems, which focuses on the physical and chemical processes involving the coastal environment.

TABLE 2.2. Key questions on the Mediterranean scale

Questions	Needed assessment
climate change-concerned questions	
How much is the near atmosphere temperature expected to increase?	Scenarios on the atmospheric circulation in the Mediterranean basin deduced by the Global Circulation Model (GCM).
	Map of expected isotherms.
What variations in precipitation are expected?	Expected new location of cyclogenesis areas.
	Map of expected isohyets.
What effects are expected on the physical conditions?	Scenarios on aridity, soil erosion and saltwater intrusion into aquifers, estuaries and other aquatic environments.
	Map of critical areas.
social change-concerned questions	
Which organisational patterns of seaport-industry interaction are expected to diffuse?	Scenarios on the main patterns of industrial seaport areas and logistic platforms which may take place.
What human pressure is expected depending on the growth of indigenous populations, migrations and tourism trends?	Scenarios on populations, migrations and tourist flows.
What changes are agriculture and crops expected to undergo as a consequence of both climate and social changes?	Scenarios on (i) hightech-based development of agriculture and crops, and (ii) conservation of traditional rural patterns.
What living marine resource uses are expected to develop?	Scenarios on fisheries and aquaculture.
Which roles will the geopolitical factors play vis-à-vis global change?	Scenarios on the inputs from possible changes in the Mediterranean co-operation.

This implies framing the individual coastal area in the global change view. To do this exercise, global change on the *global scale* is the first level reference framework. The second level consists of a view of global change on the *regional area*, such as the Mediterranean basin. The third level is the *local scale* regarding the individual area with which coastal management programmes are concerned. As focus shifts more from the global to local scale, research and increasing scenarios become more and more interesting. From the methodological point of view, the Mediterranean is a good example of the interaction established between investigations on the meso-regional and local scales.

TABLE 2.3.   A tentative agenda on the local global change

Issues	Actions
	Climate change-concerned questions
Changes in climate.	To set up scenarios on the expected variations in temperature and precipitation moving from the Mediterranean atmospheric circulation model.
Changes in sea-level.	To set up scenarios on sea-level variations and subsequent impacts on coastal areas. To set up a range of possible proactive actions needed to prevent the impacts from the expected sea-level variations on coastal areas, and to calculate the relevant costs.
Impacts from climate change on changes in water cycle.	To set up scenarios on: • intrusion of saltwater in the aquifers and estuaries; • water resource variations; • impacts from runoff on urban areas. To calculate the cost of possible damage caused by the above processes. To design possible actions to cope with the above processes, and to calculate the relevant costs.
Impacts from climate change on changes in soils and land uses.	To set up scenarios on vulnerability of soils referring to: • changes in salt balance of the soil; • changes in organic matter of soils; • fire diffusion; • desertification. *Id.*, on land degradation referring to: • rainfall and runoff erosion; • soil erodibility and erosion; • sediment budget; • freshwater quality. To taxonomise the above impacts according to their relevance to coastal management. To design a set of measures to deal with impacts, and to calculate the relevant costs.
	Social change-concerned questions
Changes in human pressure.	To set up scenarios on the expected changes in human pressure on the individual coastal areas based on population, migrations and tourism in the Mediterranean. To provide a view of the subsequent impacts on local coastal resources.
Changes in maritime transportation, seaport and industry.	To identify a range of patterns which may take place in coastal areas basing scenarios on seaport—industry systems in the Mediterranean. To design the subsequent impacts on coastal area resources and organisation from each pattern.
Changes in agriculture and crops.	To set up scenarios on possible agriculture and crop patterns and relate them to the expected impacts on the local uses from climate change. To design useful patterns to tackle these impacts.
Changes in marine living resource uses.	To set up scenarios of changes in living resource in the water column pertaining to the individual coastal areas. To focus on the possible stress to, and changes in, trophic webs. To design measures useful for the protection of the marine trophic webs from stress.

To construct ICM programmes, the questions posed in Table 2.2 may be regarded as a useful approach to the regional scale aimed at copying the local scale, i.e. taking the individual coastal area into consideration. As regards the time scale, assessment could be referred to the mid- and long-term, around 2025 and the end of the twenty-first century.

The assessment provides a reference background for coastal managers and planners concerned with a range of issues which, according to the areas, includes all or some of the issues presented in Table 2.3. Also in this case, assessment, scenarios and actions are to be referred to both the mid-and long-term.

The above framework refers to the coastal areas *per se* sketching a range of issues caused by global, climate, and social change and focusing on those measures which may be adopted to face the relevant impacts. This approach is based on the principle that the more coastal management consists of proactive measures and actions, the greater its effectiveness. Moving from this general basis, the specific framework of the core issues characterising the individual coastal area may be considered. Such a task was carried out for a range of Mediterranean coastal areas.[15] The issues and recommended actions for coastal management in Malta (Attard *et al.* 1996) are shown in Table 2.4.

TABLE 2.4. Actions dealing with impacts from climate change on the coastal area. The case of Malta

Issues	Actions
Changes in temperature	a) to examine the possibility of introducing green tarmac instead of black to lower the road surface temperature during the hot summer months;
	b) to take appropriate measures to adapt to increasing temperatures, ensuring the availability of properly air-conditioned wards and homes for both geriatric and pædiatric patients;
	c) to provide hotels and holidays buildings with air conditioning systems to meet with higher temperatures.
Changes in precipitation	a) to install a Doppler Weather Radar to provide advance warming of heavy rains, wind shear and other parameters;
	b) to develop research on dry climate technology;
	c) to introduce more drought resistant crops and trees consuming less water without loss in productivity;
	d) to establish gene banks to protect endangered species;
	e) to plant suitable trees and other appropriate plants to reduce soil erosion;
	f) to improve irrigation methods including more widespread use of drip irrigation;
	g) as precipitation may decrease and the problem of water availability may arise, to make provisions for a regular water supply, especially during the peak season in summer.
Sea-level rise	a) to set up ant put in force emergency plans to cope adequately with any eventual flooding, even if temporary;
	a) to reduce by 40 percent water production from all aquifers located at sea level;
	b) to improve drainage systems in main traffic arteries;
	c) to carry out regular maintenance of existing breakwaters;
	d) to make investigations aimed at building up new protective structures in harbours and coves around the islands in the lights of the findings on the impacts caused by sea-level rise.

A comparison of Tables 2.4 and 2.3 clearly shows how investigations become more concrete as the scale shifts from the Mediterranean to the individual coastal areas. The approach to the individual area will be more valid and successful when the former framework is further assessed and more scenarios are presented and evaluated.

CHAPTER 3

THE COASTAL ECOSYSTEM

The role of the ecosystem and the need to guarantee its integrity through adequate management approaches will be discussed. The criteria by which to provide management-oriented assessment of the ecosystem will be provided. The following questions will be considered:

Questions	Figures	Tables
Why is the ecosystem the core of ICM?	3.3	
What assessment of the plate tectonics theory is needed?	3.1	3.1
Which geomorphologic processes are to be considered as focal?	3.2	
Which are the core issues related to the marine environment?	3.2	
How may the biotic component of the coastal ecosystem be assessed?	3.3, 3.4	3.2
How may the coastal ecosystem be usefully classified?		3.3, 3.4, 3.5
What are the leading ecosystem-concerned principles of coastal management?	3.5	
What network of lines pertaining to the coastal trophic communities and their abiotic niche needs to be put on the agenda of the manager and planner?		3.6

3.1. The Coastal Ecosystem within the Coastal System

According to the sustainable development principle, the first component of the coastal system to be assessed is the coastal ecosystem. The reason for this approach is self-evident. As has been emphasised (Chapter 1.3), sustainable development is a system with three objectives: (i) integrity of the ecosystem; (ii) economic efficiency; (iii) social equity. The second and third objectives cannot be valuably pursued if the pursuit of the first is not ensured. As a result, the ecosystem has the role of the independent variable of the sustainable development equation. This postulate is the result of the double turnaround phases, which occurred between the 1972 UN Conference on the Human Environment and the 1992 UNCED, namely during the take off of the trans-industrial society.

The first turnaround phase was political. The 1972 UN Conference, which generated the environmentally-sound policy era, was basically designed to combat pollution, and so its approach was essentially concerned with physical and chemical processes. "Environment", rather than "ecosystem", was the key word. UNCED transcended that approach focusing on the ecosystem through both the Convention on Biological Diversity and Agenda 21. The second turnaround phase was due to science. Under the influence of the masterpiece by Odum[16], which designed the first approach to the ecosystem consistent with the general system theory, the organic view of Nature

51

was accepted and supported many research avenues capable of deeply influencing the management of Nature. That evolution profoundly inspired UNCED's approach, since the concept of the ecosystem was put at the heart of the political agenda and led to materials far from a determinist view of Nature.

As a result, a circular relation took shape between politics and science because both approaches were focusing on the ecosystem. A profound gap arose between those decision-makers and scientists who were following the conventional approach, considering Nature in terms of mere physical processes, namely through a mechanistic point of view, and those who were keen to transcend that view considering Nature according to non determinist approaches. That watershed, giving birth to two conflicting groups, also had implications for coastal management. There is a big difference between those managers and planners considering the environment to be a trivial machine, basically acting and evolving by force of physical and chemical processes, and those managers and planners regarding the ecosystem to be an organism capable of reacting to the same input in various ways, namely as a not-trivial machine.

To follow the innovative avenue—aimed at providing holistic views of the coastal area and conforming to Agenda 21—the concept of the *complex system*, presented in Chapter 1.6, may be adopted as the speculative background. It is essential to bear in mind that, on this basis, the approach to ICM can be carried out usefully referring to three concepts:

a) The *coastal system*, resulting from the integration of the local community and the local ecosystem and, by force of the interaction between these two components, driven towards an objective;
b) The *coastal organisation*, intended as the decision-making system, and the subsequent management processes, generating and implementing the interaction between the human community and the ecosystem;
c) The *coastal use structure*, intended as consisting of (a) the resource uses, (b) the relationships between uses; and (c) the relationships between uses and the ecosystem.

As discussed in Chapter 1.6, from this approach the coastal system is defined as a bi-modular system consisting of a *social module* (the social context and its economic organisation) and a *natural module* (the ecosystem). Its organisation is closely linked to the objective. Where sustainable development is the objective, an *ad hoc* organisation, far from the conventional organisational patterns, is needed. The present historical phase is marked by efforts of decision-making systems to pursue sustainable coastal development by adopting integration-inspired criteria. In this way, coastal organisation is stimulated to change. A long-term process has been initiated, consisting of uninterrupted re-organisation aimed at bringing the coastal system nearer to its final goal. If the language of complexity theory is used, organisation appears as the organisation of the organisation of the organisation... : an endless process operated to implement sustainability.

This Chapter will therefore be concerned with the natural module of the coastal system, i.e. the ecosystem, and the following subjects will be considered:

a) The ecosystem and its components;

b) The ecosystem's properties relevant to sustainable development;
c) The core issues of coastal ecosystem management;
d) As an example, the state of protecting Mediterranean coastal ecosystems.

3.2. The Abiotic Niche

The word "ecosystem"—a combination of the Greek words *oikos* (habitat, house) and the *systema* (collection, cluster)—was introduced in 1935 (A.G. Tansley) as a result of approaches by biology (supported by the *School of Chicago*). In general terms, this word refers to the complex of living organisms, their physical environment, the relationships between organisms, and the relationships between them and the physical environment. The *Convention on Biological Diversity* (1992) defines the ecosystem as "a dynamic complex of plant, animal and micro-organism communities and their non-living environment interacting as a functional unit" (Article 2). On this basis, it is thought of as a system including two components: (i) the *abiotic component*, constituted by the non-organic elements of the environment, which are subject to physical and chemical processes; (ii) the *biotic component*, which is subject to biological processes and interacts with the abiotic component.

3.2.1. THE ROLE OF TECTONIC PLATES

The abiotic component includes (i) the near atmosphere's elements and processes, of which climate is the main characteristic, (ii) fresh, brackish and saltwater (hydrosphere's elements and processes), as well as (iii) soil features, and geological and geomorphologic conditions (lithosphere's elements and processes). The initial step to understanding the role of the abiotic niche in the ecosystem organisation and evolution consists of focusing on the tectonic features of the continental margin within which the individual coastal area is included. According to the plate tectonics theory, designed at the mid-1960s, "the entire surface of the Earth is composed of a series of internally rigid, undeformable, but relatively thin (100–150 km) plates. Although the size of plates is variable, most of the earth's surface is covered by seven major plates (...) supplemented by several small, sometimes insignificant plates" (Kennett 1982: 131). Plates do not necessarily coincide with the continents. On the contrary, most of these are composed of both oceanic and continental crust and shift over the asthenosphere— a less rigid and more profound part of the crust—with a speed of 3 to 12 cm/yr according to the tectonic regions (Figure 3.1).

Interaction between two plates can occur in three ways. They (i) can diverge, (ii) converge or (iii) slide past each other.

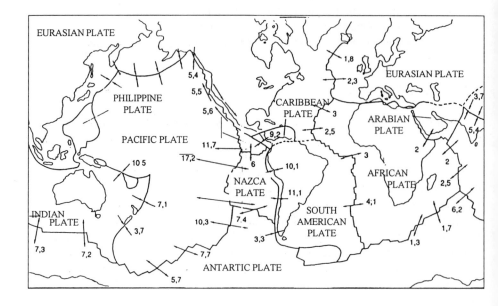

Figure 3.1. The framework of the major tectonic plates. Shift speed is indicated in cm/yr. Adapted from Vallega (1993: 32).

Divergence by two plates is marked by ridges, such as in the mid-Atlantic Ocean, the southern and south-eastern Pacific Ocean, and the southern Indian Ocean. Basalt uplifts to the oceanic surface through ridges and spread out moving from the ridge towards the peripheral area of the ocean seabed. This is the reason why divergent plates are called "constructive". The energy dynamics of ridges cause earthquakes and volcanism. Ridges are located in the deep ocean so they do not refer to coastal areas except for islands rooted on them (e.g., Iceland).

Convergence of two plates is marked by trenches where older oceanic crust is discarded and comes back to the lower layers of the crust. The convergent plates are "destructive". These plates also cause earthquakes and volcanism but, unlike the divergent plates, they concern coastal area management in vast important parts of the world, including the south- and far-eastern Asia and the islands and archipelagoes of the western Pacific.

Where two plates move parallel to each other, but at different speeds and in opposite directions along transverse fractures, crust is neither produced nor destroyed so that conservative plates exist. This motion characterises transform plates and provokes earthquakes. Transverse fractures mark the Pacific side of North America varying according to regions, and concern the coastal area.

3.2.2. THE LIFE CYCLES OF THE OCEANS AND SEAS

The origin of plates goes back to Pangæa, a proto-continent which began to break apart 245 to 208 million years ago. Since then, plates have continued to move, converging, diverging and shifting parallel to each other. Thus, oceans and seas were formed, evolved and are going to disappear. As a result, each ocean and sea is currently going through a phase of its life cycle. Each is characterised by distinct features and processes each generating different impacts on coastal areas too (Table 3.1).

TABLE 3.1.   The life cycle of ocean basins

Stage	Example	Mountains	Motions	Sediments
Embryonic	East African rift valley	block uplifts	uplift	negligible
Young	Red Sea and Gulf of Aden	block uplifts	uplift and spreading	small shelves, evaporites
Mature	Atlantic Ocean	mid-ocean ridges	spreading	great shelves
Declining	Pacific Ocean	island arc	compression	island arcs
Terminal	Mediterranean Sea	young mountains	compression and uplift	evaporites, red beds, clastic wedges
Relic scar	Indus line, Himalayas	young mountains	compression and uplift	red beds

Adapted from Kennett (1982: 179).

3.2.3. FOCUS ON THE CONTINENTAL MARGINS

Therefore, the individual coastal area can be considered by referring to the stage of life, and relevant physical processes, characterising the sea within which it is included. To carry out this approach attention focuses on the continental margins, i.e. those zones of the continental crust which took shape when a proto-continent, the Pangæa, was broken by the plate dynamics generating an ocean.[17] This makes it self-evident that the continental margins are the transition zone, usually of wide extent, *between the continental and the oceanic crust* and not necessarily coinciding with the boundaries between plates. As a consequence, the plate tectonics theory led to the understanding of the processes which have caused the margin evolution. In this theoretical framework three types of margin were identified:
a) *Divergent margins*, also called passive, aseismic, Atlantic margins;
b) *Convergent margins*, also called active, seismic, Pacific margins;
c) *Transform margins*.

Divergent Margins
"Divergent margins—Kennett (1982: 323) points out—mark the ocean–continental transition lying within a rigid lithosphere plate; that is, they are not plate boundaries. These margins develop when continents are rifted apart to form new oceans, so that the continent and adjacent ocean floor are part of the same plate. They form at diver-

gent plate boundaries initially by rifting of the continental crust and, with time, move
away from these boundaries, colliding and subsiding". They are passive and are not
influenced by volcanism and earthquakes. This tectonic setting occurs in the Atlantic
European coasts including the surrounding semienclosed and enclosed seas (from the
North Sea, Baltic Sea, etc.).

Convergent Margins
Convergent margins mark the boundaries between two different plates, such as in the
Western Pacific, where the Asian and Pacific plates converge. Unlike the divergent
margins, the continental crust and the oceanic crust belong to different plates. This is
the reason why—as has been mentioned—earthquakes and volcanism have been developed. Due to the destruction of oceanic crust occurring in these belts, the convergent
margins have always interested scientists. Together with mid-ocean ridges, they have
been the structures whose dynamics have mostly contributed in advancing the plate
tectonics theory.

Transform Margins
Transform margins, also called translation margins, consist of transform fault systems
intersecting both active and passive margins. They "result from horizontal shear motion between plates and are marked by shallow focus earthquakes. Transform margins
can become tectonically passive. During rifting, pieces of continental crust may move
relative to adjacent oceanic crust, adding to difficulties in reassembling the original
continental mass" (Kennett 1982: 325).
    When the coastal area is considered in management textbooks, the continental
margin is usually represented as structure which, moving seawards, includes three
sections: (i) the shelf, (ii) slope and (iii) rise. This is a general approach because this
three component structure does not actually characterise all the continental margins.
For example, the convergent continental margins are marked by different more complicated structures, including trenches and island arcs with intermediate seas. Besides
that, the continental slope does not mark all the margins in these marine areas. The
"classic" three components pattern of the continental margin can be seen, in its typical
features, along the Atlantic coasts where the shelf, slope and rise are well designed
and deeply influence the natural resource uses.

Continental shelves are the most important component of the margin. It is a shallow,
submerged platform whose whole extent is estimated $29x10^6$ sq km, corresponding to
about 8 percent of the entire oceanic area. They are comprised between the shore and
the shelf break, namely, the line at which the angle of inclination slopes sharply,
marking the boundary between the shelf and slope. Consisting entirely of continental
crust, shelves are terraces gently inclined seaward at an average slope of about 0.1
percent or about 2 metres per kilometre. Most shelves have a gently rolling topography. Their geology is often similar to that of the adjacent exposed portions of the
continent. The width of continental shelves varies from a few km to more than 400
km. The depth of the shelf break varies widely from about 70 m (Beaufort and

Chukchi seas in the Arctic region) to more than 400 m (off parts of Norway), with the average at about 145 m.

Continental slopes occupy about 9 percent of the sea floor (28×10^6 sq km) and range from the outer edge of the shelf to depths of 4,000 to 5,000 m. "The base of the slope is defined as the point where the sea-floor gradient drops below 1 in 40, producing a more gentle, seaward-sloping continental rise, extending from a depth of 4,000 to 6,000 m" (Kennett 1982: 321).

Continental rises, "at the base of the slopes, are immense accumulations of terrigenous sediment deposited by turbidity currents and other gravity flows and smaller quantities of pelagic sediments" (*Ib.*). They are the transition fringe between the continental and oceanic domains.

3.2.4. THE EROSION CYCLE

According to the dynamics of the water column (waves, tides and coastal currents) the coastal areas have been subject to the impacts generated by the so-called erosion cycle, a process consisting of three phases: (i) erosion, (ii) transportation, and (iii) deposition of sediment. This process leads us to distinguish two categories of coasts, namely, the *erosion-exposed coasts* and the *sheltered coasts*, the former subject to loss of rocky materials due to the action of waves and other movements of the water column, the latter benefiting from the deposition of sediment transported from erosion-subject coasts.

As far as the interaction between terrestrial areas and water column is concerned, the coastal area is thought of as consisting of two components, namely, (i) the *emerged coast*, extending up to the inland limit of the backshore, where the influence of erosion is exerted, and (ii) the *submerged coast*, extending up to the outer limit of the foreshore, where transportation and deposition of sediment occur.

Another distinction, which is also useful for biological investigations, refers to the composition of shores of either rock of soft (gravel, sand etc.) substrata. With regard to coastal topography three categories of coasts can be identified: (i) *embayed coasts*, which are rocky and characterised by inlets, bays and fjords; (ii) *plain coasts*, consisting of low-lying sand and mud areas, generally with offshore sandbars and islands; (iii) *recently-born coasts* where material has been transferred by sedimentation and accretion (Boaden and Seed 1985: 2–4).

Plain, depositional coasts are of special relevance to coastal management. They may have been subject to erosion at certain times and places due to various factors, such as storms, depletion of sediment supply. However, until the mid-twentieth century the overall, long-range tendency along these belts was that of deposition. In many parts of the world this trend is going to be jeopardised because of the sea-level variations due to climate change which was discussed in Chapter 2. Since most increasing human pressure and coastal use development concerns plain coasts it is worth specifying the distinct impacts originating from saltwater dynamics. Waves, wave-generated currents, and tides significantly influence the development of depositional landforms. In general, waves exert energy that is distributed along the coast essentially parallel to it. This is accomplished by the waves themselves, as they strike the shore, and also by the longshore currents that move along it. In contrast, tides tend to exert their influence perpendicular to the coast as they flow and ebb. As a result, the landforms taking place along some coasts are due primarily to wave processes while along other coasts they may be due mainly to tidal processes.

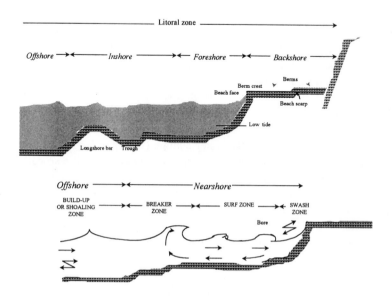

Figure 3.2. Beach profile and wave action in the nearshore area. Adapted from Kennett (1982: 298–9).

Some coasts are the result of near equal balance between tide and wave processes. As a consequence, geomorphologists speak of wave-dominated coasts, tide-dominated coasts, and mixed coasts (Kennett 1982: 287–307). *Wave-dominated coasts* are characterised by well-developed sand beaches typically formed on long barrier islands with a few widely spaced tidal inlets. *Barrier islands* tend to be narrow and rather low in elevation. Longshore transport is extensive, and the inlets are often small and unstable. Jetties are commonly placed along the inlet mouths to stabilise them and keep them open for navigation (e.g., Texas and North Carolina coasts, United States). *Tide-dominated coasts* are not as widespread as those dominated by waves. They tend to develop where tidal range is high or where wave energy is low. As a result, coastal morphology is dominated by funnel-shaped embayments and long sediment bodies oriented essentially perpendicular to the overall coastal trend. Tidal flats, salt marshes, and tidal creeks are extensive (e.g., the West German coast of the North Sea).

The Water Column

Finally, the description of the abiotic niche of the coastal area is concerned with the water column. Three kinds of water coexist in this sea-land interface:

a) *Freshwater* of the landward part of the river outlet (deltas and estuaries) and lower rivercourse, lakes and marshes;

b) *Brackish water* of the seaward part of the river outlet, lakes and marshes;

c) *Saltwater*.

The role of each component of the coastal waters needs an accurate evaluation facing the goals assigned to the coastal management programme. Nevertheless, saltwater merits special attention since most biomass exploitable by human communities depends on their physical properties and dynamics. In this respect the following aspects may be borne in mind.

Density and Temperature

The properties of saltwater vary according to the depth of the water column generating vertical stratification and are influenced by coastal currents. The water column stability, or proclivity to move vertically, depends on the change in density with depth. Besides, water temperature varies according to depth. Looking at both these properties, density and temperature, the water column may be divided into three zones: (i) a *surface shallow lens* of relatively warm water (up to 200 m depth) floating on a vast volume of much colder water; (ii) an *intermediate layer* (200 to about 1000 m) where water temperature rapidly decreases; (iii) the *lower water column* with quite uniform low temperature. The shallow and intermediate layers are separated by a zone of rapidly changing temperature, called the thermocline, and density, called pynocline. As a consequence, the amount of biomass rapidly decreases moving from the shallow to intermediate layers.

> *The Temperature of the Water Column*
> "Almost everywhere in the oceans, temperatures decrease with depth; more rapid decreases occur in the upper part of the water column. In low latitudes, typical temperature are 23°C at the surface, 8°C at 500 m, 5°C at 1000 m, and 2°C at 4000 m. There is a shallow lens of relatively warm water (to 200 m) floating on an immense volume of much colder and more saline water—the cold, deep-water zone. The surface zone comprises only about 2 percent of the ocean's volume. These cold- and warm-water spheres are separated by a zone of rapidly changing temperature, called the thermocline, and density, called the pynocline. Low-density surface water cannot easily move downward through the pynocline. The surface layer is largely a wind-driven system of currents, a turbulent zone which has little connection with deeper circulation (...). The surface layer of the oceans is well mixed by winds, waves, seasonal cooling, and salinity increase resulting from evaporation. For these reasons it is often called the mixed layer".
> (From Kennett [1982: 235–6]).

Oxygen

The near atmosphere is the primary source of oxygen in seawater and so the dissolved oxygen decreases from the surface to the intermediate depths (150 to 1,000 m) where it assumes very low values. This zone, called the oxygen minimum layer, is important because it leads to the formation of organic-rich sediments through less oxidation.

Salinity
Surface waters are marked by salinity varying remarkably according to evaporation, precipitation and other factors. It ranges from 33 per cent to 37 per cent with an average of about 35 per cent. In some seas, such as the Mediterranean and Red seas, salinity values are very high due to intense evaporation and limited water exchange with the near oceans.

Upwelling
It is not necessary to discuss waves, tides and currents. However, the importance of upwelling should be emphasised. As is well known, surface waters are subject to forcing winds. Generally winds blow parallel to the coast but, because of the associated forcing of the wind and the Coriolis effects (Earth's rotation), surface waters are pushed to move at 90° to the wind direction. As they move, cold waters ascend from the thermocline area or below to the ocean surface. As a result, the light, warm, surface water is transported away from the coast and is replaced by denser, colder deep water which is very rich in nutrients. The more the coastal area is subject to upwelling the more the marine ecosystem benefits from abundant biomass.

3.2.5. THE BENTIC AND PELAGIC REALMS

As regards the submerged coasts, the presentation of most important geological, geomorphologic and hydrological features leads us to consider two different realms: (i) the seabed and the organic life based on its sediments and physical properties, called the *bentic realm*, where the continental shelf is characterised by the prolongation of geomorphologic features of the emerged coast; (ii) the water column, including sea surface, called the *pelagic realm*.

The pelagic realm consists of *oceanic* and *neritic* provinces, the latter embracing the shallow coastal waters overlying the continental shelf. On average, it extends downward to about 200 m and, anyhow, to no deeper than 300 m. Although encompassing only little over 5 per cent of the ocean, the neritic province is highly productive so that most living resource exploitation is concentrated here. Below the neritic province the aphotic zone extends up to about 1000 m deep. As a result, when considering coastal management, attention has to be paid to the neritic province and the upper layers of the *euphotic zone*, called *mesopelagic zone*.

3.3. The Biotic Community

The biotic component of the ecosystem embraces the set of plants and animals interacting with each other and with the abiotic niche and thereby consumes energy and nutrients. In almost all ecosystems, the fundamental source of energy is radiant solar energy. Nutrients consist of the mineral and biological materials used in the metabolic processes of organisms. They are mainly composed of nitrogen and phosphorus. Both energy and nutrients decrease moving from the sea surface to the seabed so, as regards the marine coastal ecosystem, the importance of the continental shelf and euphotic zone is self evident. Regarding the ecosystem's organisation from the point of view of coastal management, a basic assessment is needed.

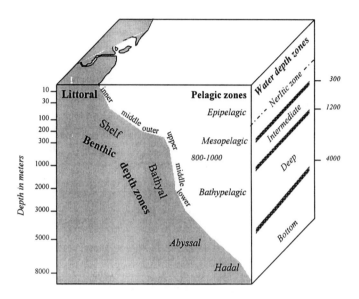

Figure 3.3. The framework of the major bentonic and pelagic environments of the marine coastal area. Adapted from Kennett (1982: 258).

   Sunlight energy is used by the ecosystem's self-sustaining, *autotrophic organisms*. Largely consisting of green vegetation, these organisms are capable of photosynthesis, i.e. they can use energy from sunlight to convert carbon dioxide and water into simple, energy-rich carbohydrates. Using carbohydrates as a basis, autotrophs produce more organisationally complex compounds, such as proteins, lipids and starches, which are essential to sustain the organisms' life processes. All the autotroph organisms give shape to the *producer level* of the ecosystem. As regards the marine coastal ecosystem, the producer level consists of plankton. Due to their limited locomotion, especially horizontally, planktonic organisms drift in response to the dynamics of the water column, especially currents. They include plants (*phytoplankton*), animals (*zooplankton*) and bacteria (*bacterioplankton*). The distinction between phytoplankton and zooplankton is not always clearly defined, since some species live as autotrophs in the light and as heterotrophs in the dark.

TABLE 3.2.   Classification of plankton by size

Type	Size range	Examples
Femtoplankton	0.02–02 μ m	Viruses
Picoplankton	0.2–2.0 μ m	Bacteria
Nanoplankton	2.0–20 μ m	small autotrophic flagellates
Microplankton	20–200 μ m	protozoans, diatoms, dinoflagellates
Mesoplankton	0.2–20 mm	Copepods
Macroplankton	2.0–20 cm	krill, arrow-worms
Megaplankton	0.2–2.0 m	large jellyfish
Adapted from: Boaden and Seed (1985: 20).		

The organic matter generated directly or indirectly by autotrophs sustains *heterotrophic organisms*. Heterotrophs form the *consumer level* of the ecosystem since they are unable to produce the food they need. Consequently, they use, rearrange, and ultimately decompose the complex organic materials built up by the autotrophs. All animals and fungi are heterotrophs, like most bacteria and many other microorganisms. As regards the marine coastal ecosystem, fish are typical consumers.

The third organisational level of the ecosystem is the *decomposer level*. Microconsumers degrade dead plants and animals and any organic matter producing minerals which are used by producers. As a result, a circular organisation sustains the ecosystem and determines its evolution.

The organisation of the ecosystem—regarded as consisting of producers, consumers and decomposers—is a general, abstract pattern. Looking at the individual ecosystem this pattern materialises in concrete organisational settings, including specific species of producers, consumers and decomposers which form a trophic web, also called biotic community. Each ecosystem has its own trophic web or set of trophic webs. A basic goal of sustainable management is to prevent human impacts from changing or destroying trophic webs. As a result, management is required to take into account not only the ecosystem's organisation *per se* but also five properties of the ecosystem.

Productivity
This property relates to the speed solar energy, a primary energy, is absorbed by the ecosystem and converted into other kinds of energy (thermic, cynetic, etc.), called secondary energy. Photosynthetic processes transform energy. The shorter the time required to move from primary to secondary energy, the higher the productivity.

Assimilative Capacity
It is intended as the capacity of the ecosystem to absorb materials into its living components via photosynthesis and digestion. This availability depends on the ecosystem's productivity.

Diversity
This property was assumed as a key word by UNCED and was the core subject of the Convention on Biological Diversity (1992). There are three ways to define diversity according to the scale on which it is considered: (i) genetic, (ii) species, and (iii) ecosystem diversity. When the organisation of the individual ecosystem is concerned— circumstance frequently occurring in coastal management-aimed approaches—*species diversity* is the most relevant concept. It is intended as "the degree of complexity of an ecosystem (...) measured in terms of the number of species and communities it contains and the degree of genetic variability within each species" (Young 1992: 35). To understand and measure the degree of diversity, species are usually divided into two categories: *dominant species*, including a great number of individuals; *rare species*, including a limited number of individuals. The greater the number of rare species, the higher the degree of diversity and the more fragile the ecosystem's organisation is.

Resilience
In a general sense, this property indicates the availability of the ecosystem to maintain its organisation in spite of being subject to disturbing inputs from the external environment. Although it is difficult to devise useful parameters to measure resilience and its evolution over time, this concept is of primary importance since it relates to the interaction of the ecosystem with other ecosystems and human communities with which it establishes contacts during its evolution. This justifies considering resilience as the most relevant property in protecting the cosystem's integrity.

Interconnectedness
This property concerns the potential or actual, capacity of the ecosystem to be linked with other contiguous ecosystems via energy, nutrients and material cycles. When this property is considered from the point of view of coastal management two geographical milieux need to be investigated: (i) the set of ecosystems included in the coastal area under management in order to understand the whole biotic organisation of the natural module of the coastal system; (ii) the set of ecosystems surrounding the ecosystem (or ecosystems) of the coastal area, in order to understand the interaction between the coastal area and its external environment. Therefore, connectivity is to be related to resilience.

3.4. The Ecosystems of the Coastal Area

One of the most intriguing subjects of ecology is the classification of ecosystems. Clearly, every adopted classifying criterion and approach has to set out to distinguish three categories of ecosystems, namely, (i) terrestrial, (ii) terrestrial-aquatic, and (ii) aquatic. The terrestrial-aquatic ecosystems embrace terrestrial-freshwater and brackish (salt-fresh water) ecosystems, and the aquatic ecosystems embrace freshwater, saltwater and brackish ecosystems. The complexity of the coastal area is due to the coexistence of a plurality of ecosystems: marine, brackish and terrestrial. The brackish ecosystems are usually the most fragile and possess the features of ecotones, i.e. the ecosystems whose structure and organisation are transitional between two distinct ecosystems. They may be defined as a "zone of transition between adjacent ecological systems, having a set of characteristics uniquely defined by space and time scales and by the strength of the interactions between adjacent ecological systems" (Risser 1990: 8). As far as the coastal area is concerned, the brackish ecosystem is transitional between the marine and terrestrial ecosystems.

Moving from this basis a wide range of ecosystem classifications may be conceived. A possible framework, partly based on the distinction between the convergent and divergent margins (Kennett 1982: 264–5), is presented in Table 3.3.

TABLE 3.3.   The coastal ecosystems

Leading edge coasts	Trailing edge coasts
Active, convergent continental margins	passive, divergent continental margins
Terrestrial ecosystems	
Forest	Savannah
Woodland	Grassland
Bush	Tundra
	coastal plain
Aquatic-terrestrial ecosystems	
Rocky coasts	Marshes
	Lagoons
	Deltas
	Estuaries
Aquatic ecosystems	
Coral reefs	salt marshes
Intertidal niche	Mangroves
Neritic niche	
Outer neritic (shelf) niche	

Bearing this framework in mind, the individual ecosystems may be considered. In this respect, the ecosystem's scale is to be assumed as the factor according to which the ecosystem can be identified and its spatial extent assessed. Like the typology of scales there is no consensus in the literature but the global (Earth as a whole), large and local scales may be assumed as reference points, at least as a preliminary approach. For example, the so-called Mediterranean Large Marine Ecosystem (MLME) can be considered as the framework for the ecosystems on the local scale. The concept of Large Marine Ecosystem (LME) is due to investigations carried out in the 1980s which led to its definition and the design of the world's map of LMEs. When this concept is adopted, attention shifts from (i) the MLME, concerned with the general ecological features of the Mediterranean, to (ii) the Mediterranean Coastal Ecosystem (MCE), focusing on the general ecological features of the Mediterranean coastal area, and to (iii) the individual Mediterranean coastal ecosystems (MCEs).

3.5.   The Large and Coastal Mediterranean Ecosystem

According to the concept adopted by the American Association for the Advanced Sciences (see Sherman and Alexander, ed., 1986), the Large Marine Ecosystem refers to an area which has unique hydrographic regimes, submarine topography and trophically dependent populations. Based on this idea, remarkable efforts in theory and investigating case studies have been made: symposia in the context of the American Association for the Advancement of Science (AAAS, 1984 and 1987), and a Conference in Monaco (1990) on *The Large Marine Ecosystem Concept and its Application to Regional Marine Resource Management* were the main scientific events. The latest event provided a more specific concept of LME:

> "Large marine ecosystems are defined as large regions of the world ocean, generally around 200,000 square kilometers, characterized by unique bathymetry, oceanography, and productivity within which marine populations have adapted reproductive, growth, and feeding strategies, and which are subject to dominant forcing functions such as pollution, human predation and oceanographic conditions. Many of these LMEs worldwide are being subject to stress from various forms of human interference or uses, including pollution and heavy exploitation of renewable and non-renewable resources against a background of growing global change."
> (From the document preparatory to the Monaco Conference).

3.5.1. THE MEDITERRANEAN LARGE MARINE ECOSYSTEM (MLME)

As defined by hydrography, the Mediterranean is a complicated LME. Complication is essentially due to two main factors, (i) the plate tectonic processes which the Mediterranean has undergone, and (ii) the biological diversity of its water column. Research has not provided incontrovertible explanation of the tectonic processes which have given shape to the Mediterranean seabed. Furthermore its origin from the ancient Tethys Ocean is controversial although a vast part of its crust and bottom morphology is oceanic. *Inter alia*, its oceanic origin is testified by the deep seabed extending down to around 4,000 m, i.e. much deeper than any other European semienclosed and enclosed sea. Although differing from each other, the models genetically explaining the Mediterranean converge to consider the evolution of this sea due to a near collision between the African and Eurasian plates. Convergence is reflected by recently-originated tectonism, volcanism and earthquakes in most of the basin, especially in its central and eastern coastal and island regions. These processes have been so intense and have assumed such different features moving from east to west that "the Mediterranean Sea is one of the most geologically complex of the world's convergent margin areas" (Kennett 1982: 392). Besides this colliding process, the Mediterranean basin has been influenced by:
a) the divergence of the Arabian plate in relation to the African plate;
b) the subduction of the African plate under the Eurasian plate;
c) the anti-clockwise rotation of the Iberian and Corsica-Sardinian micro-plates;
d) the shifting of the Turkish microplate.

As a result of these tectonic dynamics, the physiographic picture of the Mediterranean is based on three major areas:

a) the Balearic Basin, which is a vast part of the western Mediterranean and is marked by a flat, abyssal plain;
b) the Tyrrhenian (western Mediterranean) and Aegean (central Mediterranean) seas exhibiting numerous seamonts and active volcanoes giving shape to a landscape similar to that of the oceanic ridges;
c) the eastern Mediterranean province, dominated by an arcuate submarine ridge.

3.5.2. THE MEDITERRANEAN COASTAL ECOSYSTEM (MCE)

Due to the dominating collision processes most coasts of the Mediterranean are typical manifestations of the leading edge coast. They are rocky with small sand fringes. In this respect four patterns could be identified:
a) the rocky coast without gulfs and bays, as in Morocco and in some parts of Spain;
b) the rocky coast rich in gulfs and bays, as in the Ligurian and Tyrrhenian seas;
c) the rocky coast surrounded by numerous islands and archipelagoes created by the intrusion of the sea in valleys parallel to the coastline, as in Croatia;
d) the rocky coast extremely fragmented with headland, bays and gulfs and sur-rounded by numerous small islands, as in Greece and some parts of Turkey (Ionian and Aegean seas).

Typical lowlands are represented by:
a) the Libyan patterns, where desert joins the sea;
b) the Northern Adriatic patterns, where sediments have been transferred and accu-mulated by rivers (Po and others) contributing to the creation of vast lowlands in-cluding lagoons (Sestini 1992b);
c) the delta (Ebro, Po, Nile, Axios).

All these coasts are subject to erosion, in some parts accelerated by human-induced subsidence. The widest consequences are expected in lowlands (Jelgersma and Sestini 1992).

The Mediterranean Large Marine Ecosystem (MLME) is characterised by unusu-ally high biodiversity. About five hundred species of fish live in its water column, in-cluding hake, flounder, sole, turbot, sardine, anchovy, bluefin tuna, bonito, and mackerel; shellfish, corals, sponges, and seaweed are also harvested. As in the past, this ecosystem is influenced at present by the inflow of exotic species. "Many such species came through the Suez canal and are known, after the canal's builder, as Lessepsian species". For example, the Indian Ocean jellyfish was recorded in 1977 in the Mediterranean "and now forms enormous swarms in summer, from time to time" (UNEP (OCA)/MED.IG.5/Inf. 3, 1995, 58).

TABLE 3.4.   The ecosystems of the Mediterranean coastal area: terrestrial

Type	Main features	Geographical coverage	Man-made impacts
Conifer forest	Basic component: pine (*Pinus halepensis*, *Pinus pinaster*).	Higher slopes of any mountain area.	Degradation due to atmospheric pollutants and growing human pressure.
Broadleaf	Basic component: oak (*Quercus iles* in the western Mediterranean; *Quercus calliprinos* in the eastern Mediterranean).	Middle slopes of any mountain area.	Degradation due to atmospheric pollutants and growing human pressure.
Mediterranean forest	Basic components; olive (*Olea europæa oleastra*), carob (*Ceratonia siliqua*) and lentisk (*Pistacia lentiscus*).	Lower slopes of the mountain chains. In the past they were diffused in all the basin but at present they cover 9.4 percent of the terrestrial coastal area (Northern Africa, Syria and Lebanon).	In distant ages, destruction and degradation were due to the expansion of cultures and crops. In recent times, they are due to atmospheric pollutants, human settlements and tourism.
Garriga	Garriga is the result of progressive degradation of the Mediterranean forest. Species regression has included these steps: (i) degraded oak forest (*Quercus coccifera*); (ii) rosemary shrubs (*Rosmarinus officinalis*); (iii) heather moor (*Erica arborea*); (iv) asphodel moor (*Asphodelus*).	Lower slopes of many Mediterranean mountains and hills with features varying from region to region according to the species regression.	Like the Mediterranean forest.
Mediterranean maquis (shrub)	Maquis is a very rich, aromatic shrub. Main components: rockrose, laurel, myrtle, and different brooms, with olives, carob, figs, and other semitrees	Lowest slopes of the mountain chains, hills, rocky coasts of any Mediterranean region. Typical of the Ligurian and south Tyrrhenian coasts, southern France, Spain, Greece, Lebanon, Israel, Tunisia and Algeria. A Turkish variety is located in the lower slopes of Asia Minor mountains.	Degradation has increased in recent times, especially during the second half of the twentieth century. Main causes: growing human pressure, diffusion of settlements, tourism.
Mediterranean steppe	It is a transition ecosystem between the biogeographical Mediterranean zone and desert ecosystem. Basic components: great bustard (*Otis tarda*, *Chalamytodis undulata*). In the Maghreb region it is based on graminaceous, including sparto (*Stipa tenacissima*, *Lygeum spartum*).	Northern Africa, especially in Maghreb; Israel, Lebanon and Syria.	No vast degradation has been made but this ecosystem is jeopardised by the growing human pressure and settlement diffusion in the south Mediterranean.

Note: Forest ecosystems are deeply investigated by Marchand H. *et al.* (1990).

3.6.   The Coastal Mediterranean Ecosystems

Turning to coastal management attention shifts from the Mediterranean Large Marine Ecosystem to its coastal part, consisting of emerged and submerged areas. Three kinds of ecosystem therefore need to be considered: (i) coastal marine, (ii) brackish, and (iii) coastal terrestrial. To deal with this complicated framework by basically following the approach by Ramade *et al.* (1990), the *biogeographical Mediterranean zone* is assumed as the geographical coverage extending s*eawards* to the outer edge of the continental shelf, and *landwards* to the inland limit of olive tree. This zone is endowed with many types of ecosystem whose features differ according to the coastal and island regions, basically because of variable climatic, geological and geomorphologic conditions.

3.6.1. TERRESTRIAL ECOSYSTEMS

There is no consensus on the classification of terrestrial ecosystems in the Mediterranean area. Nevertheless, as a preliminary approach the framework shown in Table 3.4 may be considered useful for coastal management.

3.6.2. AQUATIC-TERRESTRIAL ECOSYSTEMS

These ecosystems have acquired different features according to the regions as regards both the biocenosis (biotic community) and landscape they bring forth. For this reason a general description cannot be provided.

TABLE 3.5.   The ecosystems of the Mediterranean coastal area: aquatic-terrestrial

Type	Location	Man-made impacts
Dunes	Northern Africa, Israel, Lebanon and Syria.	Degradation due to coastal erosion and diffusion of human settlements.
Lagoons	Venice lagoon.	Sea-level variations, pollution from atmosphere, land and sea, subsidence.
Deltas	Ebro, Po, Nile, Axios	As above
Estuaries	Any region.	Coastal erosion, pollution from river and urban settlements.
Wetlands	Located in many parts of the basin where they are called as lacs, lagoons, lidos (beach), sebkhas, bahirets.	Coastal erosion, pollution from river and urban settlements. Endangered species (plants and animals).

Note: As regards wetland ecosystems see Amanieu and Lasserre (1981); Sestini (1992a, 1992b).

3.6.3. MARINE ECOSYSTEMS

The marine ecosystems living in the continental shelf and its water column are so numerous that literature has not been able to provide a comprehensive view of all the Mediterranean, or at least of its coastal water column and seabed. This differentiation is due to many factors, *inter alia*, the rich biodiversity which has been talked about and the vast number of rare species. As a consequence, each coastal area has to be considered *per se*. The main concerns are the degradation of seabed vegetation, eutrophication and endangered species.

Posidonia Oceanica

This marine phanerogam "covers broad expanses of the inshore sea-bed, at depths ranging from 2–3 m to 30–40 m, and it is prevalent along the northern coastline in the western part of the Mediterranean. In addition to the important function they perform in the primary production and supply of oxygen to the shallow neritic zone close inshore, *Posidonia* meadows form one of the richest biocenotic communities in the demersal zone from the biological, ecological, and economic standpoints. In particular, they are feeding-grounds for many demersal fish species" (Grenon and Batisse 1989: 244). Profound degradation and vast destruction have been mainly due to man-made structures along the coasts (seaport terminals, walls, yacht anchoring, breakwaters and so-called "coastal defence structures").

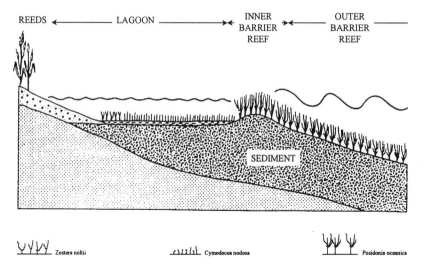

Figure 3.4. The *Posidonia oceanica* seagrass, providing the basis for the chain food of the Mediterranean marine ecosystem. From Ramade et al. (1990: 21).

Eutrophication

The Mediterranean waters are poor in nutrients, and so nutrients, notably compounds of nitrogen and phosphorus flowing to the water column from river and urban discharges, are beneficial to some extent. Nevertheless, nutrients may be discharged in quantities capable at provoking eutrophication. This anomalous process is typical of the central and northern Adriatic although it also influences many other marine areas.

"Human activities are estimated to have caused an approximately five-fold increase of phosphorus; it may only may be assumed at present that these relative increases also hold true for the Mediterranean" (UNEP (OCA)/MED.IG.5/Inf. 3, 1995, 55).

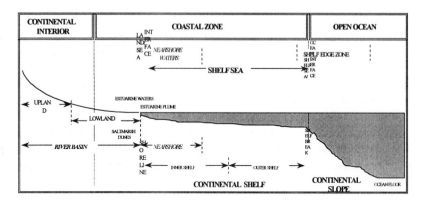

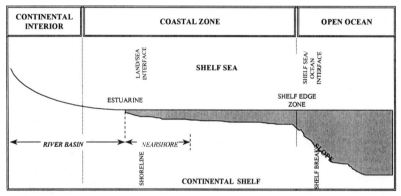

Figure 3.5. The extent of the coastal zone according to the LOICZ core project (above) and the ELOISE Science Plan (below). Adjusted from Global Change Report [1994, 33, 12] and the Ecosystem Research Report, EU, European Commission (1994: 10].

Endangered Species

Many endangered species can be found in the Mediterranean ecosystems including:
(i) flora: *Posidonia oceanica* and *cystoseira stricta* (a brown sea weed); (ii) marine fauna: monk seal (*Monachus monachus*), fin whale (*balænoptera physalus*) and common dolphin (*Delphinus delphi*); (iii) sea birds: sea gull (*Lanus audouini*), white and Dalmatian pelicans (*P. onocrotalus* and *P. crispus*), and others; (iv) reptiles: the green turtle (*Chelonia mydas*), loggerhead turtle (*Caretta c. caretta*), and others.

As has occurred in many regional seas, the Mediterranean basin has benefited from large-scale ecosystem conservation- and species protection-aimed initiatives. Apart from the programmes designed and actions undertaken on the local scale, two programmes can exert a broad influence on coastal management. While UNESCO claimed an ample number of biosphere reserves and sites as components of the common heritage of mankind (UNESCO 1990), MAP claimed one hundred coastal areas for protection (MAP/SPA 1990). Both UNESCO and MAP protected areas are being established.

3.7. Basic Questions and Key Issues

Bearing in mind both the management-oriented approach to the coastal ecosystem and the Mediterranean as a case study, coastal management clearly needs to be grounded on two basic principles:
a) the ecosystem is postulated as the main component of the resource patrimony on which the coastal community can rely to pursue sustainable development;
b) the coastal ecosystem—more specifically, the set of local coastal ecosystems—needs to be managed with holistic criteria, considering its physical and biological elements and processes, and realising that the final goal is to conserve the biotic community.

Thus seven core issues come to the fore and merit being put on the agenda of managers and planners.

The Coastal Ecosystem in the Framework of Regional Plate Dynamics

Managers and planners need a clear idea of the consequences of plate tectonic dynamics on the coastal area. Their attention ought to concentrate on the fact that *convergent margins* are usually linked with rocky coasts and can be jeopardised by seismic movements in association with volcanism. For example, this risk acquires special importance on volcanic islands and archipelagoes, such as those located in central and eastern Mediterranean waters. On the other hand, these areas are usually rich in thermal waters, on which sustainable development could usefully be based. As a result, *ad hoc* programmes are to be designed to deal with natural disasters and use natural resources efficiently.

The Relevance of the Continental Shelf and the Neritic Province

Both the shelf and the neritic zone of the water column form the most important environment for management. ICM requires:(i) exploring the risks they have undergone, or may have to undergo; (ii) designing a sustainability-inspired view of their resource endowment. In this respect, coastal erosion and relevant impacts on the biotic communities, the physical properties of coastal waters and the framework of coastal currents and other sea movements should be assessed to respond to the basis question: what may be done and should be avoided in the coastal area, mainly in the coastal water column?

The Coastal Ecosystem as a Web of Ecosystems
Effective management needs a clear view of the coexistence of terrestrial, marine and brackish ecosystems constituting the natural module of the coastal system. Their properties should be evaluated and the impacts from human presence and activities need to be assessed and measured by indicators. The organisation of the biotic community (biocenosis) of each ecosystem particularly needs to be identified with the aim of designing measures to protect trophic webs and thereby ensure that the whole community is conserved.

The Special Issues and Needs of the Individual Ecosystem
Each ecosystem is characterised by its own issues and needs, depending on the natural processes and human-induced factors. For example, many coastal areas of the northwestern Mediterranean have been subject to the destruction of the *Posidonia oceanica* meadows, which are the basis of the biocenosis of the marine trophic webs. Other areas, such as the central and northern Adriatic sea, have been liable to eutrophication, which has caused overendowment of nutrients. Where such issues have acquired significant relevance to the management of the ecosystem, specific investigations have to be carried out and special measures set up.

The Properties of the Ecosystem Facing Coastal Management
ICM only requires a subset of ecosystem's properties being investigated, possibly using indicators. This subset includes those properties—such as productivity, biodiversity and resilience—which are directly relevant to the goals of ICM programmes. Relevant measures need to be included in the programme, and actions aimed at monitoring the evolution of each ecosystem should be undertaken.

The Relevance of the Ecosystems to Sustainable Development
Two special ecosystem categories need to be identified where ICM is carried out: (i) the endangered ecosystems, and (ii) the ecosystems that merit being regarded as natural heritage for future generations. In both cases, special conservation programmes, to be framed in the coastal management programme, are needed.

Directory of Endangered Species
This is to be included in ICM programmes with measures ensuring that the single plant and animal species are protected.

3.8.   Focus on Principal Lines of Inquiry

To conclude the presentation of the ecosystem two remarks may be helpful. First, referring to the UNCED approach, clearly a much wider field of subjects is to be taken into account than in the past since the biotic components of the ecosystem are now assumed as the core of investigations and programmes. Secondly, only a selected part of knowledge provided by physical (abiotic components of the ecosystem) and biological (biotic components) sciences is relevant to ICM. As a result, the coastal manager and planners are required to select the management-relevant knowledge. Their work is typically goal-oriented.

In this Chapter the exploration of the knowledge needed, though very concise, has emphasised the importance of a set of boundaries pertaining to the ecosystem, or set of ecosystems of which the coastal area is made. Some of them were presented and discussed explicitly while others were implicit. However, they all deserve mentioning in a concluding breakdown (Table 3.6)

TABLE 3.6. The lines of coastal ecosystems relevant to management

Boundaries	Relevance
Height lines marked in maps by contour lines	
Troposphere, upper level About + 10,000 m	this line delimits the so-called near atmosphere which is subject to the greenhouse effect
Sea and land coastal birds, upper level	varying according to coastal areas, this line delimits the extent of the ecosystems' air component
Coastal forest, upper level	varying according to coastal areas and often coinciding with the watershed line, this line delimits the cover of the forest ecosystems
Coastal bush, upper level	this line delimits the lower level of coastal vegetation where grass-land and other ecosystems are also included
Depth lines marked in maps by isobath lines	
Neritic zone, lower line about -200 m	this line delimits the waters over the continental shelf and important ecosystems
Shelf edge	seawards, this line usually delimits the continental shelf and marks the transition from the shelf and slope
Slope edge about to -3,000 m	seawards, this line usually delimits the continental slope and marks the transition from the slope and rise
Rise edge	seawards, this line delimits sediment deposition and marks the transition from the coastal to deep ocean
Distance lines marked in maps by distance lines from the coastline (iso-distance lines)	
Watershed	the line delimits the land area from where fresh water is conveyed to the coastal marine waters
Backshore	Landwards from this line, the beach covered by water during storms extends
Berm crest	this line indicates the highest points of normal wave activity
Coastline	the line, varying according to short- (tides), mid- (coastal erosion) and long-term (plate dynamics) movements, is the basis on which to measure the spatial extent of ecosystems both landwards and seawards
Foreshore	seawards, this line delimits the intertidal waters
Shelf edge up to 400 km average 78 km	seawards, this line delimits the distance between the continental shelf and the coastline
Slope edge up to 600 km	seawards, this line refers to the outer edge of the slope
Rise edge up to 6,000 km	seawards, this line refers to sediment deposition and marks the transition from the coastal to deep ocean

Among others, ICM differs from the conventional approach to coastal management because of its proclivity to encompass marine spaces in programmes. The seaward limit shifts have characterised the whole history of coastal management and have had a strong impetus from Agenda 21. This justifies the special attention that literature has given to the seaward delimitation of the coastal area. Some authors have identified the seaward limit in the edge of the continental shelf; others, in the outer edge of the slope. As can be seen in Figures 3.5 and 1.2, the former approach has been adopted by LOICZ, and the latter by ELOISE.

Nevertheless, where ecology and biology are introduced in discussion, attention should undoubtedly focus on the lower limit of the neritic province, where the coastal marine life develops. This limit tends to coincide with the outer edge of the shelf. *As a result, the 200 m isobath is a significant line for ICM programmes.*

The boundaries shown in Table 3.6 do not exhaust the whole range of lines to be taken into consideration by coastal managers and planners. They only constitute a sub-set of boundary lines concerning the natural milieu. Other lines concerning the jurisdictional and administrative frameworks are essential, and they will be discussed in Chapters 4 and 6.

CHAPTER 4

ADMINISTRATIVE AREAS AND JURISDICTIONAL ZONES

This Chapter aims at focusing on how the administrative areas (land) and the jurisdictional zones (sea) are relevant to ICM and how they can be dealt with in ICM programmes. The following questions will be discussed:

Questions	Figures	Tables
How should the stewardship of the local authorities be assessed usefully?	4.1, 4.4	
How does management capability of the decision-making systems change moving from land to sea?	4.1, 4.4	
Why do the internal waters and the territorial sea provide the best jurisdictional opportunity for ICM?	4.2, 4.3, 4.4	4.1
How may optimum management of the cultural heritage be designed according to the jurisdictional maritime framework?	4.4	4.2
What is the importance of the jurisdictional continental shelf for mineral resource exploitation?	4.4, 4.5	4.3, 4.4
Why does the Exclusive Economic Zone provide the best prospect for ICM?	4.4	4.6
Why does the importance of the Exclusive Fishery Zone greatly increase?	4.4	
What methods can be applied to tailor ICM programmes to the maritime jurisdictional framework?	4.4	4.4, 4.6
What lessons may be drawn from the Mediterranean experience?		4.5, 4.7

4.1. A Double Domain

Effectiveness of the ecosystem management largely depends on the role of the concerned decision-making system. However, the role of this system and its efficiency depend on the operational sectors and geographical areas where the individual decision-making centres, giving shape to the decision-making system, operate.[18] In their turn, the operational sectors are circumscribed and geographical areas are delimited according to the national and international legal frameworks with which the decision-making centres are concerned. Hence a chain relationship: law gives shape to the decision-making apparatus; the decision making centres, interacting with each other, create management; management influences, or even determines, the evolution of the ecosystem and, *sensu lato*, that of the coastal system.

This Chapter will deal with a core subject of this chain, namely, the spatial features of the decision-making system. This consists of (i) the administrative areas, depending on national law, and (ii) the jurisdictional zones[19], referring to the law of the sea (Hayashi 1992) with special reference to the role of the jurisdictional boundaries (Prescott 1992).

75

Considering this subject, it is necessary to highlight a special boundary, called the *baseline*, whose role will be discussed in the following section. It is sufficient to note here that the baseline is different from the coastline. The former is an abstract line, claimed and designed by the coastal and archipelagic states according to the law of the sea; the latter is a concrete line separating the marine and terrestrial environments and identified by physical criteria.

The baseline separates two domains: national administration *landwards*, concerned with national law; national jurisdiction *seawards*, concerned with the law of the sea.

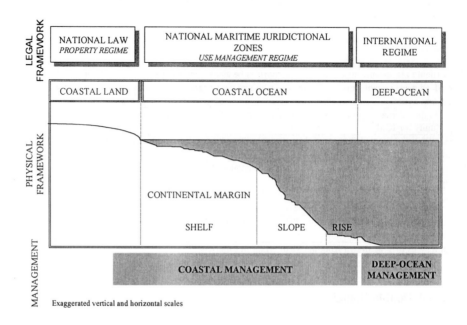

Figure 4.1. The general legal and jurisdictional frameworks relevant to ICM.

4.1.1. LANDWARDS

The space extending landwards from the baselines includes:
a) the *internal waters*, namely all those waters (salt, brackish and fresh) subject to state's sovereignty;
b) the *national land areas*;
c) the *near atmosphere* overhanging both the internal waters and national land areas.

This is the domain of the national legal system. As regards public administration (federal, state, subnational), the space consisting of the internal waters and national land areas is divided into administrative areas (municipal areas, provinces, districts, *départments*, and so on). The stewardship of governmental and local decision-making centres, such as municipalities, takes shape by planning the use of terrestrial and aquatic areas. Planning and zoning play a part in this role. Nevertheless, all plans presuppose the existence of a soil property regime. Three kinds of property can occur: (i) public (State, Crown, Federal, etc.), (ii) private, and (iii) mixed. As can be seen, the landward space from the baselines has two main basic characteristics: the direct domain of the national law and property regime.

4.1.2. SEAWARDS

Moving seawards from the baselines, first the national jurisdictional zones are crossed, then the waters, seabed and subsoil extend subject to the international regime. Coastal area management is only concerned with the national jurisdictional zones. These are maritime, since they are only concerned with the marine environment, consisting of salt and brackish waters, and their legal status is designed by the law of the sea: in practice, by international conventions. The provisions from the conventions are applied to these marine spaces only when the coastal and archipelagic states incorporate the convention in their own legal systems. This inclusion is made by ratifying the single convention, which then operates as a national law. As a result, the legal regime of the maritime jurisdictional zones is designed by the law of the sea and is operated by exerting the state's sovereignty.

The impact on coastal management from this legal mechanism is remarkable. First, the resource uses within the maritime jurisdictional zones are not legally defined by national law but by the convention regulating the individual zone. The main consequence is that the coastal and archipelagic states may only exert the rights defined by the law of the sea. Other states, called "aliens" by juridical science, have rights on the individual zones, differing from zone to zone. Therefore, coastal managers should take two spectra of rights into account: state's rights and aliens' rights.

Within the maritime jurisdictional zones property as such does not exist since this regime is replaced by the jurisdictional one. Consequently, the coastal state cannot establish its ownership on the water column, seabed and subsoil but can only regulate the uses which can be exerted in the individual jurisdictional zone. The permit regime replaces the property regime. While planning regulates property uses within internal waters and national land territory, it regulates use management within the maritime jurisdictional zone.[20]

Bearing this framework in mind, attention will be concentrated on the jurisdictional framework. Two reasons justify this approach. First, the discriminant distinguishing coastal management from conventional planning is because the marine environment is dealt jointly with the land environment to achieve integrated management. So the jurisdictional mechanism on which the management of the marine component of the coastal area is based is profoundly innovative compared with the conventional approach to planning. Secondly, human presence and resource use have expanded uninterruptedly seawards. Marine jurisdictional factors therefore play an important role.

The maritime jurisdictional zones pertaining to the coastal area will be presented considering each jurisdictional zone and focusing on what kinds of resource use are possible according to the provisions from the law of the sea. It would be useful to deal firstly with land taking the role of baselines into account, and then considering the individual jurisdictional zones exploring what rights can be exerted by the coastal or archipelagic state and what aliens' rights are recognised by the law of the sea. Once again, the Mediterranean basin will be assumed as the case study and as a ground for examples.

4.2.   The Baselines and the Internal Waters

The baselines are defined by the law of the sea. These are the dividing line between (i) the internal waters and national terrestrial spaces (landwards), and (ii) the territorial sea (seawards).

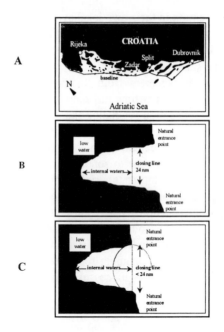

Figure 4.2. The tracing of straight baselines. Some examples: A) straight baselines in deep indented coasts (Croatia); B) juridical bay, where the distance between the natural entrance points does not exceed 24 nm; C) juridical bay, where the width of entrance points exceeds 24 nm. Adapted from Couper ed. (1983: 220).

In particular, the *United Nations Convention on the Law of the Sea* (UN LOS Convention), adopted in 1982 by the *United Nations Conference on the Law of the Sea*

(UNCLOS)[21], provided the criteria with which the baselines may be traced. Each coastal state traces its own baselines and claims them through national law.

The criteria with which the baselines can be traced form one of the most complicated subjects of international law and juridical science. As regards coastal management, the existence of two categories, normal and straight baselines, is to be considered first. *Normal baselines* are traced where a mere curvature of the coast exists. According to the LOS Convention, Article 5, where these coasts extend the baseline coincides with "the low-water line along the coast as marked on large-scale charts officially recognized by the coastal State". As a consequence, a physical geographical factor, consisting of the tide process, determines how the baselines are traced and, thereby, influences coastal management.

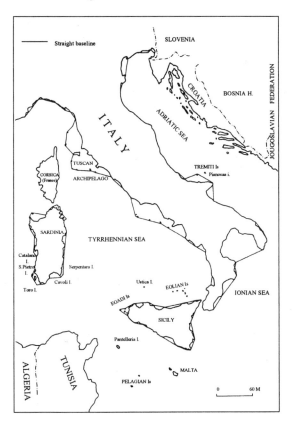

Figure 4.3. The Italian framework of normal and straight baselines. Adapted from Decree of Republic of Italy (No. 816, April 26, 1977).

Where the coast has no mere curvature, the *straight baselines* are claimed. They are regulated by the LOS Convention, Article 7: "In localities where the coastline is deeply indented and cut into, or if there is a fringe of islands along the coast in its immediate vicinity, the method of straight baselines joining appropriate points may be employed in drawing the baseline from which the breadth of the territorial sea is measured".[22] This discipline makes the straight baselines much more important for coastal management than the normal baselines. This is due to the circumstance that, where the former are traced, a maritime space of various extent is included in the internal waters and is subject to national law.

Following the management-sound approach by *The Times Atlas of the Oceans* (Couper ed. 1983: 220-1), these geographical cases are worthy of consideration.

4.2.1. DEEP INDENTED COASTS

Where coasts are deeply indented or fringed by islands, straight baselines may be drawn following the general direction of the coast. Local economic interests and the character of resultant internal waters may be appropriately taken into account. An example is the former Yugoslavia.[23] Their baselines, claimed in 1965, are 244 nm long. "The linear offshore islands represent the tops of mountain ridges which have been inundated and submerged beneath the sea (...) The northern section along the coast of Istria encloses a coast which is deeply indented. The longer section from Rt Kamenjak to Dubrovnik connects a series of fringing islands. Yugoslavia has been modest in its claim to internal waters and has left the islands called Vis, Andrija, Susac and Bisevo outside the baseline system. The baselines along this coast could well serve as a model against which other baselines connecting fringing islands could be tested" (Prescott 1985: 296).

4.2.2. JURIDICAL BAYS

Special provisions (Article 10) define the "juridical bay" as an "indentation of the coast" which is juridically regarded as a bay. The juridical bay must have an area as large as, or longer than, that of a semi-circle whose diameter is a line drawn across the mouth. In this case a closing line (straight baseline) may be drawn across the mouth between low-water marks of the natural entrance points if that distance does not exceed 24 nm (nautical miles). Where the width of entrance exceeds 24 nm, a line of that length must be drawn so as to enclose the bay. Where islands form more than one entrance to a bay, the diameter of the semi-circle is the sum total of the length of lines between the islands. Where a juridical bay is claimed, those waters extending landwards from the straight baseline are claimed as internal waters.

4.2.3. HISTORIC BAYS

A special category is that of historic bays. According to the International Court of Justice historic waters are those "waters which are treated as internal waters but which would not have that character were it not for the existence of a historic title" Frequently historical bays are of special relevance to coastal management (Francalanci and Scovazzi 1994: 82–90) of which Chesapeake Bay (USA, Delaware) is the most important in the world. In the Mediterranean two historical bays, whose claim was critically approached by juridical literature, are worthy of consideration. First, the Gulf of Sirt (or Syrtes), Libya, which was claimed as a historic bay by the Libyan Arab Jamahiriya in 1973, and whose closing line is approximately 306 nm long. Many states have joined the United States which, in 1974, protested against this claim (Prescott 1985: 298). Secondly, the Bay of Taranto, claimed by Italy in 1977, whose closing line is 61 nm long. In 1984 the United States, supported by some juridical literature, rebutted the legitimacy of that claim.

4.2.4. ARCHIPELAGIC LINES

Legally, the archipelagic state is "a State constituted wholly by one or more archipelagoes and may include other islands" (Article 46). Where these conditions occur, baselines are drawn between the outer islands enclosing the archipelagic waters. The status of archipelagic waters is recognised when the ratio of the area of water to area of land is between 1:1 and 9:1. The length of any baseline may not exceed 100 nm, although 3 per cent of them may extend up to 125 nm. As regards the Mediterranean an example of how an archipelagic state can widen the extent of its internal waters by tracing straight baselines can be found in the Maltese archipelago, where a wide marine space between the islands of Malta and Gozo—marked by important uses, including fisheries and recreation—benefits from this legal status (Francalanci, Romanò and Scovazzi 1986: 94–5).

4.2.5. UNSTABLE COASTLINES

This case occurs in lowlands and, frequently, in deltas. In tracing the baselines—LOS Convention, Article 7.2, states—"the appropriate points (connected by baselines) may be selected along the furthest seaward extent of the low-water line and, notwithstanding subsequent regression of the low-water line, the straight baselines shall remain effective until changed by the coastal State". As regards the Mediterranean this jurisdictional concern characterises the deltas of Ebro, Po, Nile and Thermaikos which—as has been mentioned in Chapter 2.3—are basically the coastal areas most subject to expected sea-level rise.

4.2.6. SEAPORT INSTALLATIONS

According to the LOS Convention, Article 11, "the outermost permanent harbour works which form an integral part of the harbour system are regarded as forming part of the coast". A significant example of how much the coastal state may widen its internal waters basing its straight baselines on the most seaward extended port infrastructure can be found in Port Said, where the baselines are connected with a very prolonged seaward entrance of the Suez Canal. Nevertheless, "offshore installations and artificial islands shall not be considered as permanent harbour works" (LOS Convention, *Ib.*). As a result, where seaports are fringed by offshore loading (such as in Maghreb) or unloading (such as in the Ligurian Sea) terminals, the coastal state cannot take benefit from this circumstance to widen its internal waters.

As already mentioned, the internal waters are legally regarded as a national space over which the coastal and archipelagic states exert all the sovereignty's prerogatives. Therefore the state has the juridical capability in this space to manage the whole coastal environment, namely, any land and marine component including the near atmospheric column above and to exploit any kind of biotic and abiotic resources. As a result, there are all the juridical conditions to design and carry out ICM.

The sea belonging to national maritime jurisdiction of the coastal state extends seawards from the baselines. According to the LOS Convention, moving seawards this space may embrace: (i) the territorial sea, (ii) the contiguous zone, (iii) the continental shelf, and (iv) the exclusive economic zone. The exclusive fishery zone and archipelagic waters may occur. It should be specified that, from the legal point of view, "continental shelf" indicates something which does not necessarily coincide with the physical continental shelf (see Chapter 3.2).

4.3.   The Territorial Sea

The territorial sea, also called "territorial waters", is particularly important for coastal management for at least two reasons. First, the jurisdictional zone is most subject to human pressure due to both human presence and resource use development. Secondly, in many states, including most Mediterranean states, this is the only existing jurisdictional zone, i.e. the only maritime space with which coastal management programmes may be concerned. It has been thought of as a prolongation of the national space, recognised by the international community.[24] In 1958, the *First United Nations Conference on the Law of the Sea* led to adopting the *Convention on the Territorial Sea and the Contiguous Zone*, which defined the concept of territorial sea: "The sovereignty of a State extends, beyond its land territory and its internal waters, to a belt of sea adjacent to its coast, described as the territorial sea". Nevertheless, the extent of these waters was not specified because the states' conference participants were unable to reach an agreement on this. In the following decades about eighty states claimed territorial seas extending 12 nm from their baselines (Treves 1983: 19). Moving from this historical process, the LOS Convention, Article 3, claimed that "every State has the right to establish the breadth of its territorial sea up to a limit not exceeding 12 nautical miles, measured from baselines (...)".

The Excessive Maritime Claims
Some states claimed their territorial seas extended more than 12 nm, giving shape to what is juridically called excessive maritime claims. Needless to say that these cases have been the subject of international controversy and many of them were submitted to the International Court of Justice. As regards the Mediterranean, Syria claimed 35 nm width territorial sea. Although this excessive claim is remarkable in itself, the most crucial issue in the Mediterranean can be found in the Aegean Sea. In accordance with the provisions from the LOS Convention regarding the delimitation of internal waters in island and archipelagic areas, Greece claimed its baselines around its islands, i.e. Lesbos, Chios, Samos and other

Dodecanese islands. As a result, the boundary of its internal waters, from which the breadth of the territorial sea is measured, was near the Turkish coastline. A big political conflict has arisen between these two countries. "Turkey takes the view that equity is the overriding principle in this disagreement. Turkey is convinced that because the Aegean islands have an area of less than 5000 square kilometres and a population of less than half a million, they cannot be entitled to a larger share of the continental shelf than accrues to the Turkish Aegean coast, with a much larger area and a population of eight million" (Prescott 1985: 307-8). Both states claimed their territorial seas as extending only 6 nm.

"While Greece would welcome an extension of this zone to 12 nautical miles it is known that in Turkey such a unilateral action is spoken of as a *casus belli*. There are three reasons why Turkey is so strongly opposed to territorial seas 12 nautical miles in the Aegean Sea. First, if both countries doubled the width of their territorial seas in the Aegean Seas, Greece's share of the waters would rise from 35 per cent to 64 per cent while Turkey's share would only grow from 9 per cent to 10 per cent. There would be a fall in the area of waters outside territorial seas from 56 per cent to 26 per cent. Secondly, except for short sections between Limnos and Lesvos, and between Khios and Ikaria, Turkish vessels would have to pass through Greek territorial waters to reach waters which were outside the regime of territorial seas. Even in these two cases the vessels would have reached waters which were totally surrounded by Greek territorial waters, so the problem of passing through Greece's territorial seas would have been only delayed if the ship was bound for the Suez Canal or the western Mediterranean Sea. Of course it would be possible for the vessels to hug the Turkish coast until they were east of Rhodes, but that is regarded as an unreasonable imposition. Turkey's third objection to increased widths for territorial seas rests on the fact that such extensions would pre-empt the question of sovereignty over some parts of the continental shelf lying 6-12 nautical miles from land".
(From Prescott J.V.R. [1985: 308]).

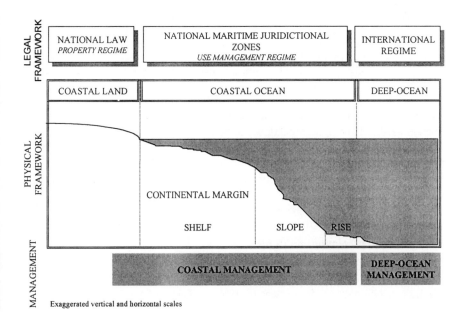

Exaggerated vertical and horizontal scales

Figure 4.4. The general framework of the national maritime jurisdictional zones. Basically, the continental shelf can be extended up to the outer edge of the continental margin or up to 200 nm from the baselines.

The importance of the territorial sea for coastal management is correlated to the range of rights of the coastal state. This range is very wide since the most important right of aliens, i.e. other states, is that of the *innocent passage*. This consists of the right recognised for merchant vessels flying other states to transit through the territorial waters provided that it is a peaceful transit, and not prejudicial to the good order or security of the coastal state. This right of innocent passage does not apply to submerged submarines or to aircraft.

The territorial sea: State's jurisdiction
LOS Convention, Article 21:
"1. The coastal State may adopt laws and regulations (...) relating to innocent passage through the territorial sea, in respect of all or any of the following:
   (a) the safety of navigation and the regulation of maritime traffic;
   (b) the protection of navigational aids and facilities and other facilities or installations;
   (c) the protection of cables and pipelines;
   (d) the conservation of the living resources of the sea;
   (e) the prevention of infringement of the fisheries laws and regulations of the coastal State;
   (f) the preservation of the environment of the coastal State and the prevention, reduction and control of pollution thereof;
   (g) marine scientific research and hydrographic surveys;
   (h) the prevention of infringement of the customs, fiscal, immigration or sanitary laws and regulations of the coastal State.
2. Such laws and regulations shall not apply to the design, construction, manning or equipment of foreign ships unless they are giving effect to generally accepted international rules or standards.
3. The coastal State shall give due publicity to all such laws and regulations.
4. Foreign ships exercising the right of innocent passage through the territorial sea shall comply with all such laws and regulations and all generally accepted international regulations relating to the prevention of collisions at sea".

Aliens do not exert other important rights in the territorial sea of the coastal and archipelagic state. This is the breakdown of their rights:
a) *Overflight*: aliens may overfly straits used for international navigation (Article 38);
b) *Fishing*: no rights;
c) *Scientific research*: research can be conducted only with the consent of the coastal state (Article 245);
d) *Submarine cables*: aliens cannot lay cables;
e) *Mining*: no rights;
f) Environmental protection: no rights.

Based on this juridical discipline, the coastal state may carry out coastal management involving all the components of the local ecosystem covered by the territorial sea. So integrated coastal management is possible (Table 4.1).

TABLE 4.1. Management capacity of the
coastal state over its territorial waters

Components of the coastal ecosystem		
Physical components	Relevant resources	
	Abiotic	Biotic
Sea surface	○	○
Water column	●	●
Seabed	●	●
Subsoil	●	●
Management capacity of the coastal state:		
○: partial management capacity; ●: whole management capacity.		

4.4. The Contiguous Zone

The contiguous zone extends seawards from the outer limit of the territorial sea. It cannot exceed 24 nm from "baselines from which the breadth of the territorial sea is measured", i.e. from 12 nm from the outer limit of the territorial sea. Not all world states have claimed their contiguous zones, as is the case of the Mediterranean (Table 4.7). On the contrary, some states, including Syria, claimed excessive contiguous zones. Within its contiguous zone, the coastal state "may exercise the control necessary to:

(a) prevent infringement of its customs, fiscal, immigration or sanitary laws and regulations within its territory or territorial sea;

(b) punish infringement of the above laws and regulations committed within its territory or territorial sea" (LOS Convention, Article 33).

As a consequence of this legal approach, the coastal and archipelagic states are only partially capable of exploiting the marine environment where the contiguous zone was claimed. According to the breakdown by Prescott (1985: 40), aliens can exercise the following wide range of rights (numbers within brackets indicate the relevant article of LOS Convention):

a) *Navigation*: they have full navigation rights providing they have not infringed regulations relating to the territorial sea (Article 35);

b) *Overflight*: they have full rights;

c) *Fishing*: the right to fish if no fishing zone, or exclusive economic zone, had been proclaimed extending beyond the seaward limit of the territorial sea; sedentary species cannot be caught (Article 77. 4):

d) *Scientific research*: they may conduct research in the water column if no fishing zone or exclusive economic zone had been proclaimed extending beyond the seaward limit of the territorial sea; no fishing can be conducted on the continental shelf underlying the contiguous zone (Article 246):

e) *Submarine cables*: they have full rights;

f) *Mining*: they have no rights;

g) *Environmental protection*: they have rights except for two limitations. First, if the contiguous zone is overlapped by an economic exclusive zone, the coastal state has authority over the seabed and the water column. Secondly, if the contiguous zone is underlined by the continental shelf, the coastal states can make regulations only for the seabed (Article 236).

As can be seen, the contiguous zone is basically concerned with the sea surface and water column. In practice, aliens have no rights on the seabed (except laying submarine cables) and the subsoil. Nevertheless, the coastal state has rights on the seabed and subsoil only if the continental shelf or the exclusive economic zone—which will be presented in the following Sections—were established. Establishing the exclusive economic zone, or fishing zone, prevents aliens from exploiting living resources, which is very important for coastal management.

As regards the coastal seas endowed with archaeological remains and other kinds of cultural heritage, the contiguous zone has special importance (Bascom 1977; Roucounas 1987; Bass 1989; Brown 1996; Couper 1996). Combining the provisions by two articles (33 and 303) of the LOS Convention it follows that, the removal of archaeological and historical objects within the contiguous zone without coastal state approval "would result in an infringement within its territory or territorial sea of the laws and regulations" (Article 303.2). Consequently, the coastal states have the right to protect and manage their marine cultural heritage. On this basis, the Mediterranean coastal and island states and local authorities may include special provisions in their coastal management programmes regarding archaeological remains and other objects of special historical and cultural importance. This is clearly relevant because the cultural heritage in many parts of the Mediterranean has a potential vital role for coastal management programmes.

TABLE 4.2.   Management capacity of the coastal state over its contiguous zone

Components of the coastal ecosystem		
Physical components	Relevant resources	
	Abiotic	Biotic
Sea surface	O	O [(a)]
Water column	O	O [(a)]
Seabed	● [(b)]	● [(b)]
Subsoil	● [(b)]	● [(b)]

Management capacity of the coastal state:
O: no exclusive capacity; ●: full management capacity.
[(a)] The coastal states have full capacity where the exclusive economic zone or exclusive fishing zone were established;
[(b)] The coastal states have full capacity where the continental shelf or exclusive economic zone were established.

4.5. The Continental Shelf

The continental shelf is the most complicated subject of the jurisdictional zones framework, basically due to the change in the approach between the First (1958) and Third (1973–1982) UN Conferences on the Law of the Sea. This jurisdictional zone was conceptually designed to regulate the interests of coastal and archipelagic states in exploring and exploiting offshore oil and gas fields. The initial definition was provided by the *Convention on the Continental Shelf* (1958), which stated that the term "continental shelf" is used as referring:

a) to the seabed and the subsoil of the submarine areas adjacent to the coast but outside the area of the territorial sea, to a depth of 200 metres or, beyond that limit, to where the depth of the superadjacent waters admits the exploitation of the natural resources of the said areas;

b) to the seabed and subsoil of similar submarine areas adjacent to the coasts of islands' (Article 1).

On the basis of that approach the *jurisdictional* continental shelf was:

a) determined according to the depth of the coastal waters neglecting the fact that the features of the *physical* continental shelf change from region to region and, therefore, it is impossible to determine *a priori* the extent of the continental shelf based on a pre-determined isobath (-200 m);

b) referred only to the seabed and subsoil, without considering the sea surface and water column;

c) regarded as a jurisdictional zone extending seaward from the 200 m isobath only where the coastal state is endowed with technologies apt to exploit oil and gas fields beyond this limit.

The last point highlights that the political motivation to the juridical concept of continental shelf was merely due to the need to meet the strategy of the offshore oil and gas industry, which was taking off at that time, and to enable any coastal state to exploit the marine environment up to the depth which made the existing technology possible. The need to tailor the juridical concept of continental shelf to the morphological features of the continental margin and to also include the other components of the marine environment (sea surface and water column) in the juridical discipline was not perceived as politically relevant.

That concept of continental shelf was replaced by the 1982 LOS Convention. According to Article 76.1 "the continental shelf of a coastal state comprises the sea-bed and subsoil of the submarine areas that extend beyond its territorial sea throughout the prolongation of the land territory to the outer edge of the continental margin, or to a distance of 200 nautical miles from the baselines from which the breadth of the territorial sea is measured where the outer edge of the continental margin does not extend up to that distance".[25]

This recently-defined approach differs from that stated in 1958 for several reasons including the following.

a) The concept of continental margin, as defined by marine geology through the theory of plate tectonics (1960s and 1970s), was introduced in the juridical approach as the main criterion to determine the extent of the continental shelf in the legal sense.

b) The depth-referred criterion to determine the extent of the legal continental shelf was abandoned, and the distance-based criterion was applied.

c) The 200 nm distance from the baselines was adopted, so following the proclivity which took place during the development of UNCLOS (1973–1982), to regard the 200 nm line as the key boundary of the maritime national jurisdiction.

d) The 200 nm criterion may be applied where the continental margin does not extend up to 200 nm from the baselines to ensure that the coastal state can benefit *in any case* from the exploitation of the seabed and subsoil extending up to 200 nm from the baselines. This means that, where the continental margin is narrow, the coastal and archipelagic states may extend their mineral resource exploitation also seawards from the outer limit of the margin, so including deep-ocean areas in their jurisdictional space.

e) The legal continental shelf refers to the seabed and subsoil extending seawards from the outer limit of the territorial sea, so confirming the conceptual approach from the 1958 Convention but explicitly assuming the 12 nm line from the baselines as the landward limit of the continental shelf. As a result, according to the LOS Convention, the 12 nm line has also acquired a vital role in managing the continental margin.

TABLE 4.3. The world framework of the jurisdictional continental shelves re mid 1940s

Delimitation criteria	Re convention	Criterion code	Number of states
Depth (200 metres) plus exploitability	1958	200m/EXP	40
Breadth (200 miles) plus continental margin	1982 Art 76.1	200/CM	22
Breadth (200 miles)	1982 Art 76.1	200	6
Exploitability	1958	EXP	5
Breadth (200 miles or 100 miles from the 2,500-metre isobath)	1982 Art 76.5	200/isobath	2
Continental margin	1982 Art 76.1	CM	1
Breadth (200/350 miles)	1982 Art 76.5	200/350	1
Breadth (200 miles) plus natural prolongation	1982 Art 76.1	200+np	1
Others (*)	—	—	2
TOTAL	—	—	80

Source: UN Division for Ocean affairs and the Law of the Sea, Office of Legal Affairs, *Law of the Sea Bulletin*.
(*) The continental shelves of Belgium and Estonia were determined according to special criteria (LOS Convention, Article 83, and coordinate-based criteria).

The LOS Convention was a profoundly innovative approach to managing the continental shelf but it further complicated the jurisdictional framework because, after that time[26], some of the coastal states continued to adopt the 1958 criterion while others began adopting the newly-defined criterion. Table 4.3 shows the present breakdown reflecting the coexistence of two different basic criteria.

The establishment of the continental shelf depends on claims and agreements. The latter procedure occurs between states with opposite or adjacent coasts where their continental shelves, if claimed by the mentioned criteria, should overlap. In this case, the 1958 Convention on the Continental Shelf, Article 6, adopted the criterion of the agreements between states. In practice, a bilateral convention is agreed to divide the marine seabed and subsoil between the two opposite or adjacent states. In the absence of such an agreement the median line between the states would have been regarded as the boundary of their continental shelves.[27] This subject was looked into the LOS Convention, Article 83.1, which stated that "the delimitation of the continental shelf (...) shall be effected by agreement on the basis of international law". The boundary between the opposite continental shelves coincides with the *median line*, consisting of the points equidistant from the nearest points on the baselines from which the breadth of the territorial sea of each state is measured. Where the coasts of states are adjacent, the boundary between their continental shelves coincides with the *lateral line*, consisting of the points equidistant from the nearest successive points on the coastline of either state. As a consequence, the median line and lateral lines have a main geographical reference for marine resource exploitation.

As already mentioned, the continental shelf was designed to enable coastal states to exploit oil and gas fields. As a consequence, the rights of the coastal state are basically limited to the seabed and subsoil. This is the framework of the aliens' rights:

a) *Navigation*: they have full rights providing they observe safety zones designated by the coastal state around artificial islands, installations and structures (Article 80).
b) *Overflight*: they have full rights.
c) *Fishing*: they have rights to fish in the water column providing that they refrain from catching sedentary species (Articles 77.4 and 78).
d) *Scientific research*: they may conduct research in the water column and, with the consent of the coastal state, also in the seabed (Articles 246 and 257).
e) *Submarine cables*: they have full rights (Article 79).
f) *Mining*: they have no rights.
g) *Environmental protection*: basically they have no rights because the coastal state has complete authority to legislate for the protection of the seabed environment (Articles 194.2 and 200).

As a result, the consequences for coastal management are those presented in Table 4.4.

TABLE 4.4. Management capacity of the coastal state over its continental shelf

Components of the coastal ecosystem		
Physical components	Relevant resources Abiotic	 Biotic
Sea surface	❍	✲
Water column	❍	❍
Seabed	●	❍
Subsoil	●	❍

Management capacity of the coastal state:
❍: no exclusive capacity; ✲: partial management capacity; ●: full management capacity.

The Mediterranean Sea has had an important role in establishing the continental shelves through agreements. This was due to a geographical factor, there being many states endowed with opposite or adjacent coasts so close that the continental shelf cannot be established through claims. The establishment of the continental shelves through agreements initiated in 1968, when Italy and Yugoslavia agreed to divide the northern and central Adriatic physical shelf into two jurisdictional shelves. Italy was induced to follow that policy as it was very interested in initiating gas exploration and production in the Adriatic waters (Bowett 1987).[28] Since then the agreements presented in Table 4.5 have been signed.

TABLE 4.5. Continental shelves of the Mediterranean Sea. Bilateral agreements

Year	States	Marine area of interest
1986	Italy—Yugoslavia	Northern and Central Adriatic Sea
1971	Italy—Tunisia	Channel of Sicily
1974	Italy—Spain	Western Mediterranean, Sea of Sardinia
1977	Greece—Italy	Ionian Sea
1984	France—Monaco	Western Mediterranean
	Libyan Arab Jamahiriya— Tunisia(1)	Central Mediterranean
1986	Libyan Arab Jamahiriya—Malta(2)	Central Mediterranean
1992	Albania—Italy	Southern Adriatic

(1) The establishment of the continental shelf was marked by this sequence of events. *1977*: special arrangement for the submission to the International Court of Justice on the difference in the design of the continental shelf boundary; *1982*: International Court of Justice, judgement; *1985*: International Court of Justice, application to revise and interpret the 1982 judgement in the case concerning the continental shelf.

(2) In 1976 a special arrangement for the submission to the International Court of Justice on the difference in the design of the continental shelf boundary was reached. In 1985 the judgement of the Court of Justice was enunciated. In 1986 the subsequent agreement between Libya and Malta was achieved.

The resulting framework shows that only a limited part of the Mediterranean, extending from the Balearic basin to the Ionian Sea, is endowed with jurisdictional continental shelves. Much of the western Mediterranean where the Iberian and Maghreb coasts facing waters rich in hydrocarbon fields, and the whole eastern Mediterranean where hydrocarbon resources are estimated to be ample, have still not benefited from the establishment of continental shelves. As regards the eastern Mediterranean, this is mainly due to the political and economic conflicts between states. In particular, the mentioned conflict between Turkey and Greece concerning the baselines and the subsequent extent of the territorial sea has been influenced by the economic interest grounded on abundant oil and gas fields. On account of the existing discipline on the jurisdictional zones, Greece is able to benefit from most subsoil resources of the Aegean Sea, even from Turkey's physical continental shelf.

4.6. The Exclusive Economic Zone and Exclusive Fishery Zone

4.6.1. THE EXCLUSIVE ECONOMIC ZONE

At present, the exclusive economic zone (EEZ) is undoubtedly the most important factor of coastal management.[29] This jurisdictional zone was designed by the LOS Convention as a result of strong political pressure from many states, including most developing countries. Before that time several states had claimed their EEZ in spite of the absence of juridical disciplines. The process continued during 1982–1994, between adopting the Convention and bringing it into force. According to R.W. Smith (1986) EEZs had been claimed by 79 states in late 1980s. In addition, taking the claims of the territorial seas and the fishery zones into account, about 28.5 million sq nm of the ocean surface were placed under these coastal states' jurisdiction.

According to the LOS Convention (Articles 55 and 57) EEZ "is an area beyond and adjacent to the territorial sea (...)" which "shall not extend beyond 200 nautical miles from the baselines from which the breadth of the territorial sea is measured". The extension of the coastal state's rights within its own EEZ is defined by Article 56.1 of the Convention:

"in the exclusive economic zone, the coastal State has:

(a) sovereign rights for the purpose of exploring and exploiting, conserving and managing the natural resources, whether living or non-living, of the waters superadjacent to the sea-bed and of the sea-bed and its subsoil, and with regard to other activities for the economic exploitation and exploration of the zone, such as the production of energy from the water, current and winds;

(b) jurisdiction as provided for in the relevant provisions of this Convention with regard to:

(i) the establishment and use of artificial islands, installations and structures;

(ii) marine scientific research;

(iii)   the protection and preservation of the marine environment".

By force of these provisions, the coastal state has similar rights within its EEZ to those pertaining to the territorial sea to the point that, as far as management is con-

cerned, EEZ could be regarded as a space functionally prolonging the territorial sea
(Table 4.6).

TABLE 4.6. Management capacity of the coastal state over its exclusive economic zone

Components of the coastal ecosystem		
Physical components	Relevant resources	
	Abiotic	Biotic
Sea surface	○	○
Water column	●	●
Seabed	●	●
Subsoil	●	●

Management capacity of the coastal state:
○: partial management capacity; ●: full management capacity.

As a result, the range of aliens' rights is very limited:
a) *Navigation*: they have full rights providing they observe safety zones designated by
    the coastal state around artificial islands, installations and structures (Articles 58
    and 60.6).
b) *Overflight*: they have full rights.
c) *Fishing*: they may have access to the surplus of allowable catch, determined by the
    coastal state, through agreements or either arrangements with the coastal state,
    which shall take into account the need to minimise economic dislocation to aliens
    who have habitually fished these waters (Article 62).
d) *Scientific research*: they may conduct research only with the consent of the coastal
    state (Article 246).
e) *Submarine cables*: they have full rights (Article 58).
f) *Mining*: they have no rights.
g) *Environmental protection*: they have no rights (Article 56.b.iii).

4.6.2. THE EXCLUSIVE FISHERY ZONE

In the mid-1990s about twenty states were endowed with exclusive fishery zones
(EFZs). This jurisdictional zone is not explicitly disciplined by the LOS Convention
and it may be regarded as the result of partially adopting the regulations on EEZ. The
coastal state may claim its own EFZ only with regard to living resource exploitation,
or it may claim that a part of its EEZ is only concerned with this use. As a consequence, the EFZ has been regarded as implicitly incorporated in the approach from
the LOS Convention to the point that the statistics from the United Nations Division
for Ocean Affairs and the Law of the Sea taxonomise it as a distinct jurisdictional
zone.

4.6.3. LESSONS FROM THE MEDITERRANEAN

When the Mediterranean is concerned in relation to these two jurisdictional zones, viz. EEZ and EFZ, a peculiar framework arises. At present, this basin is not endowed with EEZs. According to the LOS Convention, Article 74, where states have opposite or adjacent coasts the delimitation of their EEZs shall be conducted through agreements. This provision applies to the Mediterranean where the existing states are characterised by less than 400 nm between opposite and adjacent coasts. Morocco and Egypt reserved the right to claim their EEZs but this did not actually occur, and it cannot take place because the distance between their baselines and those of the opposite states is less than 400 nm. So only agreements can be adopted to establish these jurisdictional zones. In addition, in 1977 the European Commission decided that member states were not to be allowed to establish EEZs in the Mediterranean. Currently, two main reasons make it very difficult to give shape to EEZs in the basin. *First*, the geopolitical stress which, before the 1990s, was basically due to the Cold War and regional conflicts, and afterwards was caused by only regional and local conflicts. *Secondly*, the economic conflicts which affected important areas such as a large part of the Aegean Sea.

The consequences of this situation for coastal management prospects are not positive. On the one hand, the coastal states need to benefit from the establishment of EEZs in order to carry out ICM in marine areas as widely as possible and to undertake efficient, environmentally-sound, management of living resources. On the other hand, this is not possible for juridical reasons, in their turn generated by political factors. As a final result, under the MAP umbrella there is no chance in pursuing the integral protection of the marine ecosystem biodiversity. Three Mediterranean states tried to deal with this need by claiming their EFZs: Algeria (50 nm), Greece (15 nm) and Malta (25 nm) (see Table 4.7). Their claims were basically due to the need to optimise their involvement in fisheries and to mitigate competition from other countries.

4.7. Archipelagic Waters

To conclude the review of national maritime jurisdictional zones a few words should be devoted to archipelagic states. According to the LOS Convention, Article 46, "'archipelagic state' means a state constituted wholly by one or more archipelagoes and may include other islands". In this context "'archipelago' means a group of islands, including parts of islands, interconnecting waters and other natural features which are so closely interrelated that such islands, waters and other natural features form an intrinsic geographical, economic and political entity, or which historically have been regarded as such".

As has been mentioned, the archipelagic state is provided with straight archipelagic baselines joining the outermost points of the outermost islands and drying reefs (Article 47.1). Within its archipelagic waters, the archipelagic state "may draw closing lines for the delimitation of internal waters" (Article 50). Moving from the baselines, the territorial sea, contiguous zone, continental shelf, EEZ and/or EFZ may be claimed or agreed as occurs in coastal states. In the Mediterranean only Malta has the status of an archipelagic state.

4.8. The Mediterranean Jurisdictional Framework

The present Mediterranean framework of maritime jurisdictional zones is shown in Table 4.7.

TABLE 4.7.   The Mediterranean Sea. National maritime jurisdictional zones

State	CR	TS	CZ	EEZ	EFZ	CS
Albania	(*)	12	-	-	-	200m/EXP
Algeria	1996	12	-	-	50	-
Bosnia and Herzegovina	1994	12	-	-	-	200m/EXP
Cyprus	1988	12				EXP
Egypt	1983	12	24	to be determined (1)	-	200m/EXP
France (3)	1996	12	24	-	-	-
Greece (3)	1995	6/11 (5)	-	-	15	200m/EXP
Israel	(*)	12	-	-	-	EXP
Italy (3)	1995	12	-	-	-	200m/EXP
Lebanon	1995	12	-	-	-	-
Libyan Arab Jamahiriya	-	12	-	-	-	-
Malta	1993	12	24	-	25	200m/EXP
Monaco	1996	12	-	-	-	-
Morocco	-	12	24	to be determined (2)	-	-
Slovenia	1995	12	-	-	-	200m/EXP
Spain (3)	-	12	200	-	-	200m/EXP
Syrian Arab Republic	(*)	35	41	-	-	200m/EXP
Tunisia	1985	12	-	-	—	-
Turkey	(*)	6	-	-	-	-
Yugoslavia (4)	1986	12	-	-	-	200m/EXP

Notes
(1)   Egypt stated its intention to claim EEZ but, as a matter of fact, this could only be done as a result of bilateral agreements.
(2)   Morocco stated its intention to claim EEZ. It was done on the Atlantic but not Mediterranean side where bilateral agreements are required.
(3)   In 1977 the EEC decided that their Mediterranean member States should not claim, or agree on, EEZ.
(4)   Due to the collapse of Yugoslavia, its maritime boundaries are referred to Croatia, Slovenia and Bosnia Herzegovina.
(5)   The 10-mile limit applies for the purpose of regulating civil aviation.
(*)   The state did not sign the Convention.

Abbreviations
CR: Ratification of 1982 UNCLOS; *TS*, territorial sea; *CZ*, contiguous zone; *EEZ* ,exclusive economic zone; *EFZ*, exclusive fishery zone; *CS*, continental shelf; *EXP*, exploitability; *200nm/EXP*, depth (200 m) plus exploitability; *nm*, nautical mile.

Source: United Nations, Office for Ocean Affairs and the Law of the Sea, *Law of the Sea Bulletin* and http://www.un.org.

This breakdown has profoundly influenced the development of coastal management in the Mediterranean and, more important, it may well affect the prospect of disseminating ICM programmes. Clearly, the web of maritime jurisdictional zones provided by the LOS Convention is only partly reflected in the Mediterranean Sea. Only a few Mediterranean states claimed their contiguous zones (Raftopoulos 1996; The Mediterranean Institute 1987; Scovazzi 1991); a very limited number of states agreed upon establishing continental shelves; no EEZ was agreed upon and there is no current prospect of establishing them; very few EFZs were claimed. On the global scale, few EEZs have been established in semi-enclosed and enclosed seas because they often need to be delimited through agreements between opposite or adjacent states, which are politically and technically difficult to set up. However, the Mediterranean is a particular case since it is poorly endowed with jurisdictional zones and a rich endowment would be necessary to set up ICM programmes efficiently.

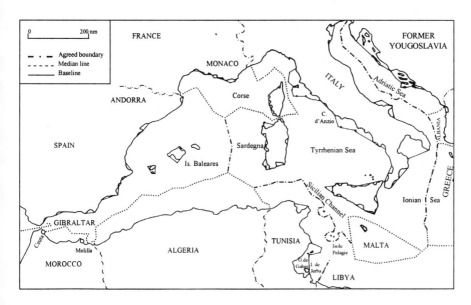

Figure 4.5. The Mediterranean jurisdictional continental shelves. Adapted from Prescott J.V.R. (1985: 302) and Vallega A. (1993: 105).

As a final result, the state's prerogatives in the Mediterranean are exerted on the territorial waters and only on limited parts of the continental shelves, where oil and gas fields have been exploited. The continental shelves in these areas are agreed according to the criteria provided by the 1958 UN Convention, based on the depth of the water column. As noted, since only a few EFZs were claimed, states have a very limited capacity to include efficient patterns of living resource use in their coastal management programmes. The impression is that, to prevent the efforts to disseminate ICM programmes from vanishing, it is necessary to provide the Mediterranean coastal belts with an endowment of maritime jurisdictional zones effectively consistent with the objectives of Agenda 21.

4.9.   Coastal Management Facing Jurisdictional Zones

Previous Sections have presented the framework of the jurisdictional zones moving from land to the sea as well as from areas subject to national law (land and internal waters) to those subject to the discipline of conventions and agreements ratified by states. The Mediterranean was presented as a useful case study in focusing on the peculiar features and key issues of the jurisdictional zone framework. As a conclusion, it should be considered how this framework ought to be taken into account when designing ICM programmes.

As a general approach, coastal planners and managers involved in the design and implementation of ICM programmes could deal with the jurisdictional zone framework effectively by adopting the following steps.

Assessing the Jurisdictional Framework
This initial step constructs a view of the existing national jurisdictional zones which are actually or potentially involved in ICM. The framework is mapped and the impacts of the individual zones on coastal management nationally are highlighted focusing on (i) the permitted and prohibited coastal uses, (ii) the resulting capacity to partially or fully manage abiotic and biotic components of the coastal ecosystem, and (iii) the capacity to protect cultural heritage.

The consistency of the existing national jurisdictional framework with sustainable coastal development and ICM prospects should be evaluated accurately. Where the jurisdictional framework does not include all the zones which could be established according to the LOS Convention, it must be assessed whether to establish new zones. This approach should lead to the design of the optimum jurisdictional framework. Critical comparison between the existing and optimum frameworks should be presented.

Focus on the Individual Zone
Attention shifts from the jurisdictional framework to the individual zones, and from the territorial sea to EEZ. The aim is to explore whether and to what extent the provisions and regulations adopted by the individual coastal state and local authorities are consistent with the need to develop ICM. For example, bearing the Mediterranean in mind, it is necessary to assess the provisions regarding navigation (sea lanes, naval areas, protected areas, etc.) in the territorial sea, as well as those regarding the protection of the marine environment against pollution from oil and gas offshore installations in the continental shelf and territorial sea. The bi- and multi-lateral agreements (general and regional conventions) are taken into account where concerned with the

individual jurisdictional zone. For example, the provisions from MARPOL are taken into account to design management patterns tailored to them. As a final result, a spectrum of provisions, consisting of international agreements ratified by the state, national law and local regulations, is framed by the coastal manager and planner in the ICM programme.

Impacts from Conflicts

The conflicts between states endowed with opposite or adjacent coasts and their impacts on the management of the individual jurisdictional zones are taken into consideration. The consequences on the prospect of building up or implementing ICM are focused on. Investigations consider whether and to what extent conflicts would be prevented by: (i) establishing new jurisdictional zones, useful for developing ICM; (ii) optimising management in the existing zones. In both cases coastal managers and planners draw up initiatives and measures to improve the framework.

The previous Chapter ended by outlining the framework of lines concerned with the abiotic and biotic components of the ecosystem and by focusing on their relevance to ICM. This Chapter has led to consider the existence of two other categories of lines, respectively consistent with the political and administrative land boundaries and the seaward limits of the national maritime jurisdictional zones. As regards the maritime context, three lines are significantly relevant to coastal management:

a) the -200 m depth line on which the boundary of most (jurisdictional) continental shelves is concentrated;
b) the 12 nm line from the baseline delimiting seawards the territorial sea;
c) the 200 nm line from the baseline delimiting seawards the EEZ and, in special cases, other jurisdictional zones.

This framework will be re-considered in depth in Chapter 6 where the optimum delimitation of the coastal area will be discussed.

CHAPTER 5

THE COASTAL ECONOMIC ORGANISATION

This Chapter aims at presenting a holistic view of the coastal economy, which implies focusing on the coastal economic organisation instead of the individual coastal sectors. It will therefore be useful to adopt methodological criteria deduced by the theory of complexity and to build up patterns of the economic organisation. To avoid a purely conceptual presentation, the Mediterranean will be considered as a way of example. The following questions will be discussed:

Questions	Figures	Tables
Why should the coastal economic organisation be assessed instead of the coastal economic structure being assumed as a leading concept?	5.1	
What methodological approach may be adopted to assess coastal economic organisation?	5.1	5.1
By what criteria may the leading components of coastal economic organisation be identified?		5.2
How may the large geographical space be taken into account to identify the economic organisational models of the individual coastal area?		5.3
How may the Economic Organisation Model of the Coastal Area of the individual coastal area be designed?	5.5	5.3
What methodological issues should be dealt with in focusing on coastal economic organisation?	5.1	

5.1.   A Conceptual Pathway

The coastal ecosystem, forming the natural component of the coastal system, was presented in Chapter 3. Attention now shifts to the social module of the system and focuses on the organisation of the coastal area. As already discussed[30], the concept of organisation has a vital role facing the building up and implementation of the integrated coastal management programmes. This is worth clarifying with the following chain of concepts.

- The coastal system is composed of two modules, the coastal ecosystem (natural module) and the socio-economic context (social module) interacting with each other and with their external environment.
- The conventional approach to the coastal area implies investigating the social module focusing on the economic sectors on which the coastal system is based and providing distinct views of each sector (seaports, navigation, industry, fishing, agriculture, tourism, etc.). This is the product of a structuralism-based view of coastal reality.

99

- Holistic approach, deriving from the theory of complexity, differs from the conventional one because it requires global views of the coastal system to be designed with the final aim of building up an integrated coastal management programme. There is a close feed-back between the holistic approach, concerned with science, and integration, concerned with action: the more the former is methodologically correct, the more the latter is effective and encourages the adoption of holistic methods.
- The setting up of global views of the social module requires assessing the interaction between the coastal resource uses to be provided so leaving out the mere sectoral description.
- The result of concentrating on the interaction between the coastal resource uses leads to designing organisational models. For example, when the interaction between the seaport and littoral manufacturing plants processing imported raw materials is focused on, the model of a maritime industrial development area may be designed. This exercise marked geographical investigations on cityports during the 1970s (Vigarié 1981).
- Moving from interaction between coastal resource uses, the whole *economic organisation model of the coastal area* (EOMCA) may be designed to provide a holistic view of the coastal economy.

If the reasoning based on the above chain of concepts is accepted, discussion could focus on how a useful method for designing EOCMAs may be adopted. This issue will be dealt with in the following sections, first methodologically before applied to the Mediterranean.

5.2.   A Methodological Route

An economic organisational model of the coastal area, and of any regional space may be well designed in a variety of fairly sophisticated ways which are generally supported by quantitative investigations. So a wide spectrum of possible methods may be envisaged and the following methodological steps provide a simplification.

a) *The time scale*. Research should concentrate on the second half of the twentieth century since the economic and social processes which have influenced the evolution of the coastal areas, as well as the targets to which they could lead, mainly originated during those decades.

b) *The large geographical space of reference*. At present, the economic evolution of the individual coastal areas is deeply influenced by the ways in which they interact with other regions. As a result, the ICAM programme is more effective when there is more focus on the large geographical space within which interaction has taken place. For instance, the ICAM programme of a coastal area of Tunisia or Croatia may not be designed if the interaction of these areas with the Mediterranean context is not taken into account. Consequently, the large geographical scale of reference needs to be determined as a preliminary approach to the design of EOMCAs.

c) *Leading process on the large geographical scale*. The processes which have marked the evolution of the large geographical area are identified and their influence on coastal regions is circumscribed.

d) *Distinctive economic organisational model*. It is postulated that the influence exerted by the individual processes, which have marked the evolution of the large geographical space of reference, have brought about or implemented the ways of using the coastal resources. The final product of the interaction between the inputs from the large space and the approach to the resource uses is the rise and implementation of distinctive local organisational model. For example, the tendency to increase the domestic production of oil and gas, which has been implemented in the Mediterranean in the 1970s, led to the creation of offshore industrial areas marked by the interaction of navigation, manufacturing plants, seaports, and other sectors.

e) *Economic organisational model of the coastal area*. As already mentioned, the association of distinctive economic organisational models gives shape to a model representing the whole organisation of the coastal area. It could be assimilated in an organisational model of a region, and imagined as the product of a rectangular matrix where:

- *in lines*, the distinctive organisational models are indicated;
- *in columns*, the features of the EOMCA, resulting from the association of the distinctive model, are presented.

The above methodology will be applied to the Mediterranean coastal areas in the following sections. In particular, the distinctive economic organisational models (Section 5.5) and the resulting EOMCAs of the Mediterranean coastal areas (Section 5.6) will be considered.

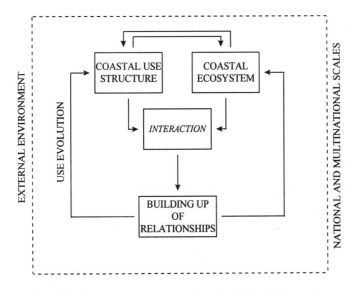

Figure 5.1. The methodological approach to the economic organisation of the coastal area.

5.3.   The Time and Geographical Scales of Reference

The Mediterranean coastal areas should be regarded as significant case studies of the evolution which the world's coastal areas have undergone during the latter half of the twentieth century for at least three reasons. First, these areas have been subject to enormous growing human pressure causing serious concerns with regard to harmonising resource uses and protecting the ecosystem. Second, the economic organisation of many of these areas has been based on various associated activities (seaports, tourism, industry, fishing) bringing about vast and intensive exploitation of coastal resources. Third, developed and developing coastal areas have coexisted and interacted even under geopolitical stress involving the Mediterranean basin.

The evolution of the Mediterranean economic context may be regarded as a case study positively testing the stage-based historical view presented in Chapter 2.2. In fact, the Mediterranean was the geographical area which triggered the transition from the neo- to the trans-industrial stage in the early 1970s as a result of the fourth Arab-Israeli war and collapse of the conventional organisation of the international oil market. Hence, the historical approach shown in Table 5.1 may be usefully adopted.

TABLE 5.1.   Economic development.
A stage-based historical view

Stage	Timing
Neo-industrial: take off	1900–1950
Neo-industrial: maturity	1950–1970
Neo-industrial: decline	1070s
Trans-industrial: take off	1970–1990
Trans-industrial: maturity	Post 1990

Attention may be concentrated on the maturity and declining phases of the neo-industrial stage and the take off and the maturity of the trans-industrial stage. As has been mentioned, evolution and changes can be considered from the geopolitical, economic and social points of view. Useful elements for framing the organisation of the individual coastal area in the Mediterranean are concisely given in Table 5.2.

5.4.   Leading Processes and Economic Models: A General View

Moving from this approach to the time and geographical scales, the economic evolution of the Mediterranean may be regarded as deeply influenced by four processes:
a) increase of urban pressure with the subsequent rise of megalopolises;
b) industrial development linked to the import of minerals and energy sources (northern side of the basin) and the processing of domestic resources (southern side);
c) rise and diffusion of transport nodes based on seaports, coastal cities and airports;
d) expansion of land and marine exploitation of living and cultural resources.

Table 5.2. Global change: economic and social components. The Mediterranean as a case study

Neo-industrial stage Maturity phase: 1950-1970	Trans-industrial stage Take off phase: 1970-1990 Maturity phase: 1990s onwards
Geopolitical features and processes	
World Cold war development. Decolonisation. Establishment of multi-national areas for economic co-operation and political alliance.	World End of cold war. Political regional stress. Diffusion of regional wars. Decline of stability of the former Soviet Union. Establishment of new multi-national areas for economic co-operation.
Mediterranean East-west political and military confrontation. Conflicting policies between Arab and Western countries. Second (1856) and third (1967) Israel-Arab wars. Decolonisation. Terrorism. Closing of the Suez Canal (1967-1975).	Mediterranean Slow peace process in the Middle East. Stress between Arab countries. Stress in the former Yugoslavia. Cooperation-based process generated by the UNEP Mediterranean Action Plan. The building up of a Mediterranean policy by EU (1995).
Economic features and processes	
World End of the Bretton Woods-based policy, initiated in 1944. Subsequent liberalisation of financial markets. Regionalisation of economies on a multi-national basis. Changes in the international division of labour due to the rise of the developing world and the tertiary-oriented evolution of developed economies. Rise of the unitisation and standardisation in transport systems. Rise and diffusion of environmentally-sound policies.	World Creation of the world financial market. Globalisation of world economy. Profound changes in industrial and tertiary organisation because of information and communication techniques. Birth of Internet system. Increasing gap between developed and developing worlds. Fragmentation of the developing world into areas characterised by increasingly different growth levels. Adoption of the sustainable development principle by the international community (Rio Conference).
Mediterranean Development of Arab countries due to oil and gas field exploitation. Expansion of the European Economic Community (EEC) Mediterranean coverage and co-operation treaties between EEC and southern Mediterranean countries. Decreasing of primary industry-based growth in the western European side; subsequent growth and diffusion of this sector in the southern side. South-North migrations.	Mediterranean Decline of economic growth of Arab countries. Development of Turkey. Establishment of the European Union (UE) (1993) and subsequent enlargement of co-operation both eastwards (Former Yugoslavia) and southwards (Mashrek and Maghreb). Development of sustainable development-based policy by the member states of the Barcelona Convention. South-North and East-West migrations. Creation of a Mediterranean tourist network based on maritime transportation (cruise and tourist circuits).
Social features and processes	
World Increasing gap between developed and developing countries as regards social conditions. Social stress caused by terrorism.	World Recognition of the human development concept. Focus on health, food, women and children. UN Conferences on social equity. Adoption of the principle of rights of future generations. In spite of the conceptual progress and design of human development-sound policies, worsening of social conditions of both the developed world (especially urbanised areas) and developing world. Adoption of concept of the cultural heritage. Efforts to protect endangered communities (small islands, indigenous peoples or rain forests). Efforts to protect children.
Mediterranean Extended social stress brought about by regional wars and terrorism. Discrimination of minor groups, women and children in the eastern and southern sides.	Mediterranean No adoption of social policy by member states of the Barcelona Convention nor by EU. Weakening of efforts to protect cultural heritage in the Mediterranean basin. Expanded social stress in the eastern side (former Yugoslavia), Mashrek (Palestinians) and Maghreb (Libya). Increasing role of Integralism and subsequent stress.

5.5.   Leading Processes and Economic Models: A Detailed View

These processes have provoked or implemented many kinds of interaction between economic activities, giving shape to a vast range of distinctive economic organisational models. A concise view may be found in Table 5.3 while the mentioned Mediterranean processes and subsequent organisational patterns are presented in this Section.

TABLE 5.3.   Leading Mediterranean economic processes
and subsequent distinctive local organisational patterns

Leading economic processes	Distinctive local organisational patterns
Urban pressure	UP1—Linear megalopolis
	UP2—Small- and medium sized urban settlement
Tourism development	TD1—Traditional tourist settlement
	TD2—Modern tourist settlement
	TD3—Recreational harbour
	TD4—Marine recreational plant
Industrial development	IA1—Import-dependent heavy industrial area
	IA2—Export-oriented heavy industrial area
	IA3—Industrial-technological pole
	IA4—Logistic platform
	IA5—Offshore industrial area
Node and gateway diffusion	NG1—Bulk-dependent seaport terminal
	NG2—Industry-serving seaport terminal
	NG3—Containerised seaport terminal
	NG4—Passenger seaport terminal
	NG5—Fishing harbour
	NG6—Multi-function seaport
Biomass exploitation: land	LB1—Self-sustaining land biomass use
	LB2—Conventional market-oriented land biomass use
	LB3—Hightech and market-oriented land biomass use
Biomass exploitation: sea	MB1—Catching-based marine living resource use
	MB2—Farming-based marine living resource use
Natural resource exploitation	NP1—Integrally protected area
	NP2—Special purpose protected area
Cultural resource exploitation	CP1—Historical land site
	CP2—Historical waterfront
	CP3—Land archaeological area
	CP4—Marine archaeological area

5.5.1. URBAN PRESSURE

Urbanisation, international migrations, together with growing and expanding tourism and recreational uses, have been the main factors causing increased human pressure in the Mediterranean. Since 1980, the population has increased from 84.5 to 123.7 million inhabitants.[31] All these factors have exerted profound impacts on many Mediterranean coastal regions and islands. At this stage, two basic models concerned with *Human Pressure* (UP) may be identified.

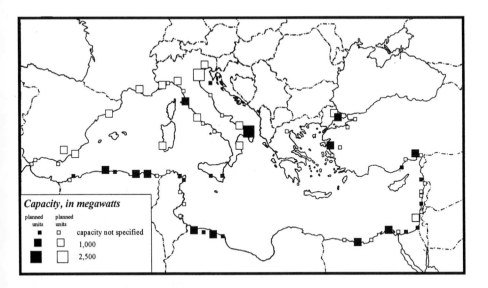

Figure 5.2. Thermal power-stations of the Mediterranean coastal areas. Adapted from Grenon and Batisse (1989: 139).

UP1—Linear Megalopolis

This model was described by *ekistics theory* according to which the world's urbanisation tends to increase uninterruptedly. Small megalopolises consist of multi-cities and multi-town urbanised areas of no more than 25 million inhabitants. They have diffused in the Mediterranean extending along the coasts and inland valleys. Significant examples can be found in the north-western Mediterranean—where a small megalopolis extends from the north Tyrrhenian Sea (Leghorn) to the Gulf of Lion (Marseilles)— and in the terrestrial area extending on the opposite sides of Bosporus. The creation and growth of small megalopolises in the EU side has been associated with increasing migrations from Central and Eastern European, African and Asian regions, so that the social contexts have become increasingly multi-ethnic.

UP2—Small- and Medium-Sized Urban Settlement

This is the traditional model of Mediterranean coastal and island settlements. It consists of close small- and medium-sized towns located along the continental and island coasts. During the 1970s and 1980s this pattern declined almost anywhere in the EU side because of growing coastal urbanisation and the creation of linear megalopolises where many small towns have been encapsulated. It has persisted in many other parts of the basin, such as the Dalmatian and Greek coasts, the Mashrek and Maghreb coasts and on many islands, including large ones, such as Rhodes, Malta, Sicily, Sardinia and Corsica.

5.5.2. TOURISM DEVELOPMENT

Tourism is a main component of Mediterranean economic growth, and is expected to exert increasing influence on coastal landscape and cultural heritage management. At the present time four patterns of settlements and man-made structures characterising *Tourism Development* (TD) can be found.

TD1—Traditional Tourist Settlement
These settlements mark historical towns and small coastal and island centres where tourism was initiated at the begin of the twentieth century. They consist of small quarters, ancient hotels and other kinds of settlement. Frequently settlements are incorporated in the pre-existing urban texture giving shape to a harmonious landscape. All the Mediterranean coastal belts—from the Gulf of Naples to the Thassaloniki Gulf, from the Turkish to the Tunisian coasts—are more or less endowed with towns including historical tourist settlements.

TD2—Modern Tourist Settlement
Modernism has involved the Mediterranean with the diffusion of tourist villages and other similar settlements, such as "Club Méditerranée". They possess four features closely concerned with coastal management, since: (i) they are generally located quite far from existing urban settlements and their success depends on maintaining this spatial separation; (ii) they are characterised by profound cultural extraneity *vis-à-vis* historical settlements and landscape; (iii) they can damage the natural values and cultural endowment of the environment due to their proclivity to be located in attractive sites; (iv) in many cases, most added value from this activity flows outside the Mediterranean because these settlements are managed by multi-national companies.

TD3—Recreational Harbour
Due to the expansion of yachting and sailing throughout the Mediterranean, recreational harbours have diffused. Two basic patterns have characterised this process. First, many recreational harbours have been created within the existing merchant seaports and, as a consequence, they are closely linked to urban settlements surrounding the port facilities. This type of harbour may be found all over the basin, from the Côte d'Azur to the Turkish coasts and it frequently serves sailing vessels. Secondly, there are increasing number of recreational harbour, consisting of specialised structures for both sailing and motor vessels, which provide many complementary facilities, including hotels and services for vessels and navigation. Many of the modern harbours, especially on the African and Asian sides of the basin, are located in coastal places far from cities and, like tourist villages, are significant signs.

TD4—Marine Recreational Plant
Coastal and marine parks, extending in the brackish belt, and submarine parks, accessible through submarine vessels, and fixed plants and facilities located on the seabed, converge to create the new frontier of Mediterranean tourism. As land and marine landscapes draw more and more tourists, these installations will spread making the urban and industrialised coast less attractive.

5.5.3. INDUSTRIAL DEVELOPMENT

During the 1980s marine and transport geography carried out many investigations into industrial areas which were then diffusing along the coasts of developed countries, from Japan to Western Europe. Those areas were built up to process imported energy sources and minerals to supply the domestic market. The evolution of that model led to create generations of the so-called Maritime Industrial Development Areas (MIDAs), each of them marked by its own organisation. Three generations were identified by Vigarié (1981): the first was found in developed countries and consisted of areas designed to process imported raw materials without caring for the environment; the second differed from the first because the location of plants and their organisation were designed also with the aim of protecting the environment and human health; the third was similar to the first and has taken place in the developing world.

Those areas slightly declined between the second half of the 1980s and early 1990s. The subsequent changes have caused great differentiation both in terms of economic and spatial processes. At the present time five models of *Industrial Areas* (IAs) in the Mediterranean can be identified.

IA1—Import-Dependent Heavy Industrial Area
This model consists of plants processing imported energy sources (usually oil and coal) and minerals to supply the domestic electric network (Grenon M. *et al.* 1993), as well as iron and steel, petrochemical and other plants conveying semi-finished goods to the domestic market. The needs of these plants are met by specialised liquid and solid bulk port terminals and, in some cases, even by offshore terminals. This pattern has diffused along the EU side of the Mediterranean, in some regions of the extra-EU side, such as Kastela Bay (Croatia), and can be found in selected areas of the Asian side, such as Izmir (Turkey). There are frequently strong environmental impacts. Since the late 1970s manufacturing areas have undergone a decline in the EU coastal regions. The most important example of this organisation can be seen in the Fos area (Gulf of Lion).

IA2—Export-Oriented Heavy Industrial Area
This kind of industrialisation is based on heavy industries processing local raw materials to supply the domestic market and, to a large extent, to export semi-finished goods. Also these plants intensely interact with the local seaport where onshore and offshore loading terminals were built up. Such areas have diffused along the Asian and African sides. As regards the latter, Annaba and Sfax (Tunisia) may be regarded

as significant examples. These areas are a main consequence of the new inter-regional division of labour which has solidified during the take off phase (1970–1990) of the trans-industrial stage.

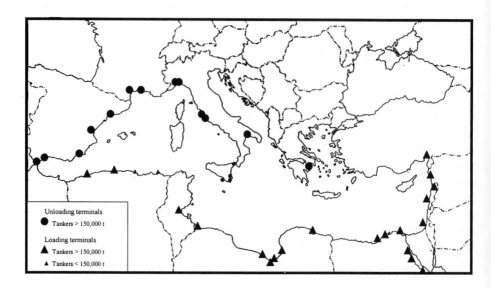

Figure 5.3. Tanker loading and unloading terminals in the Mediterranean. Re: late 1980s. Adapted from Grenon and Batisse (1989: 173).

IA3—Industrial-Technological Pole

This coastal model has arisen in the EU regions, in some cases as a result of the decline and subsequent conversion of heavy industry-based areas. It consists of industries not heavily dependent on the importation of energy sources and minerals and producing finished, frequently hightech, products. Association and, frequently, close integration between the manufacturing plants and local institutes and laboratories carrying out technological research, especially in the field of computer science and communication technologies, are the main characteristics. According to the development of the individual area this model can be subdivided into two sub-models: on the one hand, the pole essentially based on manufacturing facilities; on the other essentially based on research.

IA4—Logistic Platform

As containerised traffic has expanded along intra- and trans-Mediterranean routes the need to build up logistic platforms—such as those of Fos, Genoa Voltri, Gioia Tauro and Haifa—has diffused. Logistic platforms were designed not only for loading and unloading containers but also to manage the final operations of the manufacturing provide facilities for commercial operations, such as consignment. This structure, which is closely related to market globalisation, is also expected to diffuse in the Mediterranean.

IA5—Offshore Industrial Area

This type of industrialisation consists of oil and gas exploration and exploitation platforms. The more numerous they are, the deeper they influence other uses of coastal waters and the more they bring about environmental impacts. This model is limited to the Adriatic and Ionian seas and a few other parts of the western (Balearic Sea) and central (Libyan waters) Mediterranean.

5.5.4. NODE AND GATEWAY DEVELOPMENT

The Mediterranean has an unprecedented number of seaports. Except for a few seaports, such as Fos-Marseilles, this endowment is marked by medium- and small-sized seaports along the continental and island coasts of the whole basin. Their relevance to coastal management is twofold. On the one hand, they are located in important coastal areas, such as bays and gulfs, and have provoked environmental impacts both on the ecosystem's abiotic components, such as the alteration of seabed morphology, and the biotic components, such as the destruction of seabed grasslands. On the other hand, they have deeply interacted with manufacturing and tourism facilities and they have acquired the role of gateways for the regional organisation. Where this process has taken shape, fairly expanded networks of relationships have been created between the coastal area and inland regions. The greater this interaction, the greater is the prospect of stimulating development. As a result, there being either a node for traffic and a gateway for a region has marked the role of a well developed seaport city. Six types of *Nodes and Gateways* (NG) can be identified in the Mediterranean.

NG1—Bulk-Dependent Seaport Terminal

This structure diffused during the maturity phase of the neo-industrial stage (from the mid-1950s to the mid-1970s) along the EU Mediterranean coasts. More recently it has expanded in the African and Asian sides. In the EU side unloading terminals prevail conveying bulk to inland areas; the southern Mediterranean sides are characterised by loading terminals and bulk for export.

NG2—Industry-Serving Seaport Terminal

This structure is similar to the above but seaport terminals convey energy sources and minerals to local, coastal manufacturing plants. As has been mentioned when industrial coastal organisation was discussed, this kind of model has diffused along the EU coasts during the neo-industrial stage, and afterwards on the southern sides.

NG3—Containerised Seaport Terminal

As regards containerised traffic, two spatial structures can be found. First, the full containerised seaport, where only containers are handled by lift-on lift-off and roll on-roll off handling systems. Characteristic features of this model have taken shape in Algeciras (Spain), Genoa Voltri and Gioia Tauro (Italy). Secondly, there are the part-containerised seaports, where containerised terminals coexist together with bulk and other kinds of terminals. This model is very diffused. The most significant examples are Barcelona (Spain), Fos (France) and Piraeus (Greece).

NG4—Passenger Seaport Terminal
Because of the development of passenger transportation, mainly due to cruise and high-speed vessels, many Mediterranean seaports have been restored and have enlarged their historical passenger facilities including "maritime stations". In the meantime other seaports have benefited from newly-built facilities. As a result, during the take off of the trans-industrial stage (1970s and 1980s) the Mediterranean has been endowed with a constellation of these facilities, the development of which is expected to continue, at least in the short- and mid-term.

NG5—Fishing Harbour
Just a few seaports are only serving fisheries, while many of them have installed facilities for this sector consisting of unloading terminals, conservation and forwarding plants associated with facilities for merchant vessels.

NG6—Multi-Function Seaport
Most Mediterranean seaports are multi-functional due to the coexistence of bulk facilities, passenger and containerised facilities and, in some cases, facilities for fisheries.

5.5.5. LAND AND MARINE BIOMASS EXPLOITATION

Shifting to the exploitation of biotic components of the coastal ecosystem both the land and marine uses of biomass are to be considered. A detailed classification may be sketched based on the sectors of activity—forest, farming, agriculture, fishing, and others. However, when coastal management is the reference point, three models of *Land Biomass* (LB) use and two patterns of *Marine Biomass* (MB) use may be taken into consideration.

LB1—Self-Sustaining Land Biomass Use
Exploitation of forest, shrub, olive- and vine-growing, and horticulture have remained the economic bases of many coastal and island communities, especially in the southern and eastern Mediterranean. Nevertheless, this model, typically self-sustaining, whose roots go back to the most ancient civilisations of the basin, is expected to survive only in limited parts of the basin due to the development of urban settlements and mass tourism. This affects also the rights of future generations because, in many coastal and island regions, technology and social expressions of this traditional exploitation of land biomass are core components of the local cultural heritage.

LB2—Conventional Market-Oriented Land Biomass Use
In this model, olive- and wine-growing, horticulture and fruit-farming are practised to convey the subsequent products to the extra-regional markets. However, both techniques and commercial organisation remain based on conventional criteria. This model characterises many areas, from Tunisia and Lebanon to Dalmatia and Turkey, but it is expected to have restricted geographical coverage because of the diffusion of hightech-based activities, urbanisation and mass tourism.

LB3—Hightech Market-Oriented Land Biomass Use
This kind of coastal resource use marks those forest and agricultural areas which benefit from bio-engineering and computer science assistance, hightech water supply, co-operation from research bodies, advanced commercial facilities, and others, to the point of being able to convey their products to the European and extra-European markets. The horticulture-based areas of the Valencia and Murcia regions, as well as Israel coastal areas, are significant examples of this high value added pattern.

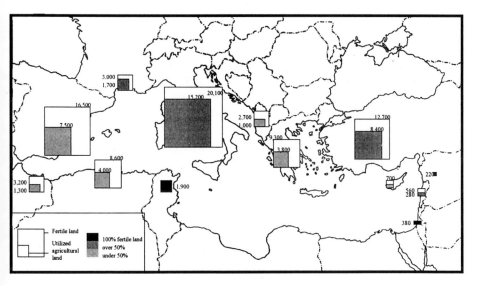

Figure 5.4. Surface of fertile land and utilised agricultural land in the Mediterranean catchment area (1,000 hectares). The larger squares indicate the surface area of fertile land, i.e. soil without major factors limiting fertility, in the Mediterranean catchment area of each country. The smaller squares indicate the utilised agricultural land in the areas. The shading of smaller squares suggests the intensity of pressure on fertile lands in the more or less accentuated tendency to utilise lands in the Mediterranean region. Adapted from Grenon and Batisse (1989: 82).

MB1—Catching-Based Marine Living Resource Use
This traditional model of living marine exploitation is rooted on traditional fishing techniques, some of which, such as trawl-net fishing, may well be abandoned (Demetropoulos 1981; Brunel *et al.* 1987; Charbonnier D. *et al.* 1990).

MB2—Farming-Based Marine Living Resource Use
This model, based on aquaculture techniques, is spreading out from developed to developing regions and is expected to expand rapidly. At the present time, the most important areas are located in Spain, France and Italy.

5.5.6. NATURAL AND CULTURAL HERITAGE EXPLOITATION

During the take off of the trans-industrial stage (1970s and 1980s) both the local ecosystem and cultural heritage have been perceived by local communities and decision-making centres as increasingly important for economic development and quality of life. This process is strengthening as far as the sustainable development concept is adopted by decision-making centres as the focal goal of their strategies, and is perceived as essential by local communities. As a result, coastal managers and planners are catalysed by the attractive prospect of regarding the ecosystem and cultural heritage as resources for development. In this respect two models of *Natural Protected* (NP) areas and four models of *Cultural Protected* (CP) areas may be identified.

NP1—Integrally Protected Area
This model may be referred to land and marine parks and other kinds of protected areas where all the biotic and abiotic components of the ecosystem remain integral. Conservation is pursued to its full extent.

NP2—Special Purpose Protected Area
This kind of area occurs where only one component, or group of components, of the local ecosystem is protected. This is the case of coastal and island natural reserves where a selected number of endangered species of plants and animals are protected, and to some extent human settlements are permitted.

CP1—Historical Land Site
Historical settlements are located in all the Mediterranean coastal and island areas and form the main attraction of small islands. These are basic components of coastal management.

CP2—Historical Waterfront
Many Mediterranean cityports are endowed with urban historical waterfronts. The more waterfront revitalisation diffuses in the Mediterranean, the more these sites will acquire a leading role for coastal organisation and will serve as basic resources for sustainable development. As a result, their importance for coastal management and planning is expected to increase.

CP3—Land Archaeological Area
These areas are a main components of the cultural heritage of the Mediterranean coastal areas and islands. Their protection is a prerequisite for the sustainable coastal development.

CP4—Marine Archaeological Area
The progress of techniques to discover marine archaeological remains, to gain access to them and to use them for underwater tourist circuits is widely regarded as a culturally-sound use of the coastal area. The Mediterranean has an unprecedented patrimony of marine archaeological remains, most of which are well known the world over. As a result, the management of the marine archaeological heritage is expected to become a core component of the organisation of many Mediterranean coastal and island areas.

5.6.   The Economic Organisation Model of the Coastal Area

The distinctive economic organisational models presented in the previous Section form the basis on which the Economic Organisation Model of the Coastal Area (EOMCA) can be constructed. This final passage of the methodological pathway proposed in Section 5.2 to design a holistic view of the economic organisation of the coastal area can take place by considering each coastal area to be marked by one or more distinctive economic organisational models. So, by considering which models are associated within the individual coastal area, the whole organisational model of the area may be designed. When the Mediterranean is taken into consideration, this approach leads to the drawing list of typical EOMCAs presented in Table 5.4.

As can be seen, this table consists of a matrix where:
- *in lines*, the distinctive economic organisational models are presented;
- *in columns*, the Coastal Organisation Models (EOMCAs) which have take place as a product of the association of two or more distinctive economic organisational models, are presented.

The individual EOMCA provides a holistic view of the whole organisation of the coastal area, presenting those features and processes relevant when ICM programmes are to be set up. As an example, Table 5.4 includes some EOMCAs which have marked the economic growth of coastal uses and their diffusion in developed and developing Mediterranean regions, and with a potential strong influence on the design of ecosystem- and cultural heritage-sensitive management models. This breakdown is the result of a preliminary approach. Further investigations could lead to a different framework. What is under discussion here is the methodological concern. In this respect, the main role of Table 5.4 is that of providing room to set up useful methodologies. The basic question is whether and to what extent this approach is useful for designing those views which are needed when coastal management models are to be set up consistently with the guidelines provided by Agenda 21, Chapter 17, and subsequent prescriptions and recommendations, such as UNEP's.[32] Moving from the breakdown presented in this matrix the coastal manager can proceed along two steps.

TABLE 5.4. Coastal organisation models. The Mediterranean

Disctintive economic organisation models	Economic Organisation Model of the Coastal Areas (EOMCAs)						
	Big city-based developed area	Big city-based developing area	Small and medium cities-based developed area	Small and medium cities-based developing area	Urbanised recreation-based area	Mass modern tourism-sustained area	Integrated agriculture and industry area
UP1 Linear megalopolis	●						
UP2 Small- and medium-sized urban settlement		●				●	●
TD1 Traditional tourist settlement			●	●			
TD2 Modern tourist settlement			●		●	●	
TD3 Recreational harbour			●		●	●	
TD4 Marine recreational plant					●	●	
IA1 Import-based heavy industrial area	●			●			●
IA2 Export-oriented heavy industrial area	●	●					
IA3 Industrial-technological pole	●						
IA4 Logistic platform	●						
IA5 Offshore industrial area			●				
NG1 Bulk-depending seaport terminal	●	●		●			●
NG2 Industry-serving seaport terminal	●	●					
NG3 Containerised seaport terminal	●	●					
NG4 Passenger seaport terminal	●						
NG5 Fishing harbour				●			●
NG6 Multi-function seaport	●	●	●				●
LB1 Self-sustaining land biomass use				●			
LB2 Conventional market-oriented land biomass use	●		●				● ●
LB3 Hightech market-oriented land biomass use	●						
MB1 Catching-based marine living resource use			●		●	●	
MB2 Farming-based marine living resource use					● ●		
NP1 Integrally protected areas					●		
NP2 Special purpose protected area					●	●	
CP1 Historical land site		●	●	●			●
CP2 Historical waterfront		●	●		●		
CP3 Land archaeological area			●	●			
CP4 Marine archaeological area			●				●
Examples	*Barcelona*	*Izmir Bay*	*Antalya*	*Kastela Bay*	*Côte d'Azur*	*Costa Smeralda*	*Syrian Coast*

At the present time the only example of coastal areas based on expanded offshore oil and gas exploration and exploitation can be found in Ravenna (Italy).

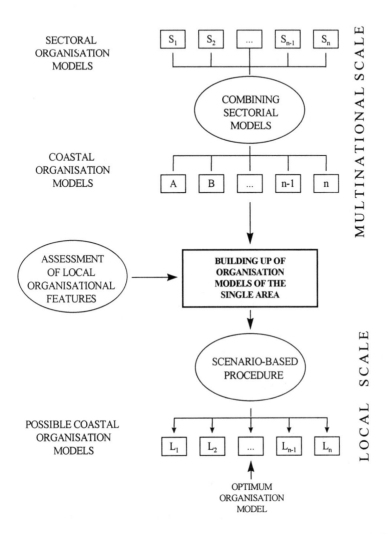

Figure 5.5. The methodological approach to the building up of the Coastal Organisation Model (COM).

The *first step* leads to the building up of subsequent, more specific matrices. This implies shifting from the regional (multi-national) scale, which in our case is from the Mediterranean as a whole, to the local scale, concerned with the individual coastal area. In this case a second, matrix, closely derived from the first one, would be constructed where:

• in *lines*, the coastal organisation model marking the individual coastal area is presented with its components, consisting of the distinctive economic models;

- in *columns*, the impacts on the ecosystem, social context and cultural heritage are presented.

For example, in the southern side of the Mediterranean numerous coastal areas have been involved in the diffusion of *mass tourism*. Moving from EOMCA shown in Table 5.4, the organisation model of these areas may be presented through a matrix including:

- *lines*, which encompass: recreational harbours, revitalised waterfronts, tourist residential settlements, sunbathing and diving areas, market centres, aquaria, museums, etc.;
- *columns*, which encompass: land space use, beach use, fresh water use, acceleration of coastal erosion due to man-made coastal structures, decline of traditional activities, rise of new professional sectors, impacts on historical settlements, archaeological remains, indigenous traditions, etc.

The *second step* consists of the design of possible future coastal organisation models. This scenario-based exercise is essential to design and operate adequate approaches methodologically. They may be conducted moving from the organisation models sketched in the first step, which reflects the present state of coastal resource uses, and taking into account the possible inputs arising from the external environment and indigenous process in the short- and mid-term. For example, the exercise, carried out by the Blue Plan of MAP to design the "possible futures" of the Mediterranean, could be usefully adapted to construct the possible organisation models by which the individual coastal area could be marked.

To clarify the features and possible results of the methodological route presented in this Chapter, an application to the Antalya coastal area, located in the southern side of Turkey, is presented in the Appendix to this Chapter.

5.7. Key Issues and Basic Options

The approach presented in this Chapter is quite different from those usually adopted by coastal area management programmes. In this respect, some basic differences from the conventional and innovative approaches are worth pointing out to make the coastal manager and planner aware of the possible options when they intend to undertake or optimise an ICM programme.

The conventional approach, widely experienced in the field over the last twenty years, usually focuses on the so-called key issues marking the coastal area under investigation. For example, this has led to designing coastal management programmes concentrated on tourism growth (*developmental prospect*) or coastal erosion (*environmental issue*). The present approach differs from the conventional one because it focuses on the coastal organisation to provide a holistic view of the interaction between the individual resource uses, and the associated use of various resources, and to consider the resulting whole spectrum of consequences. In epistemological language, the conventional approach is structuralism-oriented and uses disjunctive logic while the present one is complexity-referred and uses conjunctive logic.[33]

The conventional approach, which is widely reflected in the 1975-1995 MAP experience, considers the coastal economic features and sectors *per se* while the present approach attributes primary importance to the external environment with which the individual coastal area interacts. In order to do that, the whole Mediterranean has been approached not starting from the coastal area in itself but from the space from which the individual coastal area and reacts to impulses. According to this approach, investigation of the individual coastal area aims at ascertaining whether and to what extent its organisational model is consistent with the corresponding model identified by considering the Mediterranean as a whole. The assessment is always referential. For example, as can be seen by the Appendix to this Chapter, the model of *small and medium cities-based developed area* of Antalya (Turkey) would be built up and analysed by relating it to the corresponding general model concerned with the Mediterranean basin which is presented in Table 5.4. Thus, the coastal manager and planner commutes between the large geographical scale model and that of the individual coastal area to compare features and processes and design scenarios. As a result, the assessment of the coastal organisation is complex system-oriented, based on identifying set of elements and processes relevant to the building up of integrated management, and methodologically antipodal to the conventional exhaustivism-inspired descriptions having value *per se*.

The conventional approach looks at the future of the coastal area according to projective methods. It implies the future being regarded as a prolongation of the past. The innovative approach consists of looking at the future according to prospective, scenario-based approaches. As a result, a range of possible economic organisation model of the coastal area (EOMCAs) may be designed and the optimum model among them, i.e. the most consistent with the sustainable development principle, may be identified.

There is no doubt that the innovative approach is more difficult to operate than the conventional one. Furthermore, the use of computer science tools, such as the coastal Geographical Information Systems (GIS), is more problematic. The coastal manager and planner should evaluate this basic option to be aware of the different outcomes they may achieve according to the conceptual and methodological routes they pass through.

Appendix

THE ECONOMIC ORGANISATION OF ANTALYA COASTAL AREA

Introduction
This Appendix has the aim of concisely presenting an example of how the methodo-
logical approach, discussed in Chapter 5, may be applied. After emphasising some of
the main aspects in this area, the distinctive organisational models will be considered.
Then, there will be the discussion of how the Economic Organisation Model of the
Coastal Area (EOMCA) will be constructed, how scenarios may be built up, and the
results usefully encapsulated in the integrated coastal management (ICM) programme.
Finally, the interaction between the models of economic organisation concerned with
the large geographical area and the individual coastal area will be dealt with.

Antalya: A Presentation
Facing the homonymous Gulf (Antalya Körfezi), on the Mediterranean side of Turkey, Antalya, with 250,000 inhabitants, is one of the most important resort on the
Turkish Riviera. Its cultural heritage is one of the richest in the country including the
old town surrounded by fortified walls, an ancient tower, a mosque dating from 1250,
a Byzantine church converted into a mosque, and an archaeological museum. It is located in a plain surrounded by the Taurus mountains and enjoys a subtropical warm
climate. Its coastal area, in its turn rich in archaeological remains and historical settlements, produces cotton, sesame, citrus fruits, early vegetables and bananas. Inland
areas are marked by grain and livestock. Antalya is a significant example of the intense rapid tourism development if the Mediterranean region of Turkey. High stan-
dard hotels, new villages near the old and modern city, shops, an old port converted
into a recreational harbour, navigation, fishing and sunbathing facilities have constituted a promising basis for the tourism development.

Phase 1—The Distinctive Economic Models
Due to the coexistence of important natural and cultural resorts and its location in the
Mediterranean waters between the isles of Rhodes and Cyprus, the economic endowment of Antalya includes several groups of activities. They form economic modules
closely related to the distinctive economic models presented in Table 5.4. This is the
relevant breakdown where there are also the main kinds of resources on which the development processes are based.

Phase 2—The Economic Organisation Model of Antalya Coastal Area
This breakdown leads us to cluster the Antalya coastal area into the *Small and medium
cities-dependent developed area*, presented in Table 5.4. This pattern has diffused in
all the Mediterranean and may be regarded as leading to the development of the
coastal areas. As can be seen, the above methodological step had two functions. On
the one hand, it has led to define the economic organisations of the Antalya coastal
area to be based on nine groups of functions, concerned with settlements, cultural

heritage, recreational uses and living marine resource uses. Each function, e.g. recreational harbour, consists of various types of activities (vessels, maintenance, supply, etc.), which may be identified by analysing it in detail. On the other hand, relating the Antalya model to the Mediterranean *Small and medium cities-dependent developed area*, it is possible to evaluate how the former is reflected in the latter, which is a model with more general features suitable for a large geographical space. So doing, the Mediterranean model may be implemented according to the results of investigating the Antalya coastal area. As a consequence, there is a feed-back: the Mediterranean model helps to provide a holistic view of the Antalya coastal area; the view of the local coastal reality leads to implementing the Mediterranean model; and so on.

TABLE 5.5. The Antalya coastal area: distinctive economic organisation models

Distinctive Economic Organisation Models (Re: Table 5.4)	Resources for development processes			
	Abiotic natural resources (1)	Biotic natural resources (2)	Cultural resources (3)	Landscape (4)
UP2 Small- and medium-sized residential settlement	☐		☐	
TD2 Modern tourist settlement	☐			☐
TD3 Recreational harbour	☐			☐
NG5 Fishing harbour	☐	☐		
LB2 Conventional market-oriented land biomass use		☐		
MB1 Catching-based marine living resource use		☐		
CP1 Historical land site	☐		☐	☐
CP2 Historical waterfront	☐		☐	☐
CP3 Land archaeological area			☐	☐
CP3 Marine archaeological area			☐	☐

(1)   Space, water, building materials;
(2)   Forest, land and marine living resources, crops;
(3)   Archaeological remains, historical building, old town, museums;
(4)   As a visual resort.

Phase 3—The Impacts on the Ecosystem and Cultural Heritage
The breakdown presented in Table 5.5 provides the basis for assessing the impacts that both the local ecosystem and cultural heritage of the Antalya coastal area have undergone on account of the economic organisation. Investigations may lead to the design of a breakdown where rows represent the distinctive economic organisational model of the Antalya coastal area, and columns show the impacts. As an example, the web of impacts is presented in Table 5.6.

TABLE 5.6.   The Antalya coastal area: impacts from the economic organisation

Distinctive Economic Organisation Models	Impacts							
	SP	PB	IA	CE	HS	LE	ME	SD
Small- and medium-sized residential settlements	O	O	O	O	O	O	O	O
Modern tourist settlements	O	O	O	O	O	O	O	O
Recreational harbour					O	O	O	O
Fishing harbour						O	O	
Conventional market-oriented land biomass use						O		
Catching-based marine living resource use							O	
Historical land site								
Historical waterfront								
Land archaeological area								
Marine archaeological area								

Codes: SP, human Settlements' Pressure on land space use; PB, human Pressure on Beaches; IA, risk of salt water Intrusion into Aquifers; CE, risk of accelerating Coastal Erosion; HS, risk of degradating Historical Settlements; LE, Land Ecosystem degradation due to human pressure; ME, Marine Ecosystem degradation due to navigation and land sources; SD Scenic beauty Degradation because of new settlements and facilities.

The above breakdown has been designed by way of example. It could acquire different features where accurate investigations are carried out on site. Moreover, it serves as ground to provide more detailed views by desegregating the distinctive individual economic models in their components. For example, the model *Fishing harbour* may include various components tentatively presented in Table 5.7.

TABLE 5.7.   The Antalya coastal area: possible impacts from fishing harbour

Distinctive Economic Organisation Models	Impacts							
	SP	PB	IA	CE	HS	LE	ME	SD
Berthing facilities		O		O	O	O	O	O
Vessel supply		O			O	O	O	O
Vessel repair		O			O	O	O	O
Refuelling facilities		O			O	O	O	O
Conservation plants					O	O	O	O
Market facilities								
Fishermen houses	O				O	O		O
Bar and restaurants					O	O		

Phase 4—Scenarios on the Economic Organisation of the Antalya Coastal Area
The possible developments of the coastal organisation according to the objectives pursued by the local decision-making systems are designed and investigated. For example, as in many Mediterranean coastal areas subject to tourism pressure and endowed with rich cultural patrimony, the Antalya development will depend on three basic factors, namely on how and to what extent
a) mass tourism will be encouraged;

b) the land and marine local ecosystem will be conserved;
c) the land and marine cultural heritage will be conserved;
d) the land and marine scenic values will be protected.

All these factors lead to the design of various scenarios: mass tourism development; culturally- and environmentally-sound tourism; expanded residential facilities; limited residential facilities; conservation of the natural heritage through the establishment of protected areas; development based on the intense use of natural heritage; integral conservation of cultural heritage; partial conservation of cultural heritage; conservation of coastal agriculture or its decline because of increasing residential and tourist land use; and so on.

Each scenario may be presented following the methodological route adopted to assess the present economic organisation that has marked Phases 2 and 3. In other words, each scenario will be presented by identifying (i) the distinctive economic models by which it is marked, (ii) the abiotic, biotic, cultural and scenic resources on which it is based, and (iii) the environmental, human and cultural impacts which may be generated.

Phase 5—The Design of the Optimum Economic Organisation Model
The range of scenarios built up in Phase 4 is evaluated according to the principle of sustainable development. In this way the consistency of each scenario in protecting the ecosystem's integrity, economic efficiency and social equity is considered and discussed. This approach leads to the design of the optimum economic organisational model of the coastal area.

CHAPTER 6

THE GEOGRAPHICAL EXTENT OF THE COASTAL AREA

This Chapter will focus on how the coastal area may be delimited by adopting criteria consistent with the integrated management principle and the concept of sustainable development. The following questions will be discussed:

Questions	Figures	Tables
Why is the geographical delimitation of the coastal area a key issue?	6.1	6.1
How was this issue initially approached by states?	6.1	6.1
What contexts may be taken into account?	6.2, 6.3, 6.4, 6.5	
Which land and marine reference boundaries are to delimit the coastal area?	6.2, 6.3, 6.4, 6.5	6.3
Which delimitation criteria merit being adopted?		6.3
What is the role of the continental shelf and margin?	6.5	

6.1. Holism and Organisation: Scientific and Political Key Words

The coastal area is a complex spatial system also because its spatial extent—in height, depth and surface—is a basic component of its organisation. No manager or planner could deal with the coastal area without considering the role of its geographical three-dimensional extent. How the optimum delimitation of the coastal area could be designed in order to optimise the pursuit of the management objective is not a new question. It has been much discussed in the literature, as well as by the technical materials provided by intergovernmental organisations. An example of the former approach is the attention paid by Sorensen and McCreary (1990) to this subject *vis-à-vis* the design of the optimum legal and institutional arrangement framework for coastal management. An example of the latter approach was presented by the UNEP guidelines (1995) on integrated coastal management. However, the question of delimitation has acquired increasing, and perhaps unexpected importance as a consequence of the principles stated and guidelines provided by the UNCED Agenda 21. Such an ample consideration is due to at least two main factors. First, as a result of inputs from Agenda 21, the focus of coastal management was set on protecting the ecosystem's integrity and so the delimitation of the coastal ecosystem, or that of a set of coastal ecosystems, has become a main subject. This was the reason why attention focused on the *ecosystem's boundaries* in Chapter 3. Secondly, coastal area management has been regarded as the product of progressive involvement of the coastal sea in management

123

patterns. This was why the administrative boundaries (on land) and the jurisdictional limits (at sea) were taken into consideration in Chapter 4.

At this point those *mises au point* on the ecosystem and the legal and jurisdictional framework must be recalled to respond to the recurring question when methodologies have been needed to design and carry out ICM programmes: what is the optimum geographical delimitation of the coastal area? In other words, the question is *what is the perfect geographical extent of the coastal area where integrated management is pursued*.

The complexity of the question is due to two concurrent circumstances. First, focusing on the ecosystem, the perfect geographical area should encompass a single local ecosystem, or set of contiguous ecosystems. Secondly, this geographical coverage requires being tailored to the spatial stewardship of local decision-making systems. The more the extent of stewardship coincides with that of the ecosystem the more ICM is effective. This statement leads to the deduction that the coastal area is a special complex system: *complex*, because it can be assessed and managed only by holistic criteria; *special*, because it is characterised by special concerns generated by the need to tailor management to the local ecosystem, or a group of contiguous ecosystems.

Before proceeding along this avenue, it is necessary to note that the subject has been dealt with by different approaches according to the geographical scale on which is has been considered. Where the national scale is taken into account, the objective is usually to provide general guidelines within which the individual coastal management programmes are to be framed. Where the local coastal management programme is considered, the coastal area needs to be designed moving from the national guidelines, where they exist, with the aim of meeting the specific needs and objectives of the local decision-making system and coastal community. The subject becomes much more concrete and the national guidelines on coastal area management have the role of serving as the general framework *vis-à-vis* the local context.

6.2.   The Conventional Design of the Coastal Area

The first deep discussion on this subject seems to date back to the early 1980s when the United Nations (United Nations 1982) emphasised its relevance to the need of optimising the approach to coastal management. Since then, the delimitation criteria of the coastal zone (national scale) or the coastal area (local scale) have acquired different approaches according to the concepts of coastal area that have been adopted by, or that were buried in the minds of managers and planners. The criteria through which the coastal area could be geographically delimited in accordance with the integrated management principle may usefully be defined if the way to deal with this subject has been approached during the last quarter of a century is borne in mind.

During the 1970s and 1980s, when environmental policy was deeply influenced by the United Nations Conference on the Human Environment (1972), numerous criteria were applied to spatially delimit the coastal area for management purposes. The national laws and provisions by the USA Coastal Zone Management Act (1972) provided a wide range of criteria which may be placed into four categories: (i) arbitrary criteria; (ii) criteria based on the physical coastal features; (iii) criteria concerned with economic organisation, or with (iv) juridical and jurisdictional factors. According to Sorensen and McCreary (1990: 10), the framework presented in Table 6.1 characterised the late 1980s.

TABLE 6.1.   Land and maritime boundaries of national programmes on coastal management

Country and states	Coastal zone boundaries	
	Landwards	Seawards
Brazil	2 km from mht	12 km from mht
California		
- 1972–1976 planning	highest level of nearest mountain range	3 nm from the cb
- 1972–1976 regulation	1,000 yds from mht	3 nm from the cb
Costa Rica	200 m from mht	mlt
China	10 k from mht	15 m isobath (or depth)
Ecuador	variable line according to the issues of the individual area	-
Israel	1–2 km depending on resources and environment	500 m from mlt
South Africa	1,000 m from mht	-
South Australia	100 m from mht	3 nm from the cb
Queensland	400 m from mht	3 nm from the cb
Spain	500 m from highest storm or tide line	12 nm (outer limit of the territorial waters)
Sri Lanka	300 m from mht	2 km from mlt
Washington state		
- planning	inland boundaries of coastal counties	3 nm from the cb
- regulation	200 ft from mht	

Adapted from Sorensen and McCreary (1990: 10).
Acronyms and abbreviations: *cb*, coastal baseline; *mht*, mean high tide; *mlt*, mean low tide; *nm*, nautical mile; *ft*, feet; *yds*, yard.

Arbitrary Criteria

These were mainly grounded on the concept of distance. As an example, in some countries the landward boundaries of the coastal area vary between 300 and 500 m from the mean high tide, and the seaward boundary is identified as the 3 nm line beyond the coastline.

Physical Criteria

Table 6.1 shows that, as regards national policy, the mean high tide and, in a few cases, the mean low tide, have been the most frequently used physical criteria to delimit the coastal area. Meanwhile, the proclivity to adopt other criteria, especially landward, to design the individual coastal areas for management purposes has grown. As regards the landward extent, the watershed between rivers flowing to the sea and those flowing to inland was regarded as relevant. Also the landward limit of the back-shore has been taken into consideration. The outer edge of the continental shelf was regarded as the optimum seaward limit (Oregon).

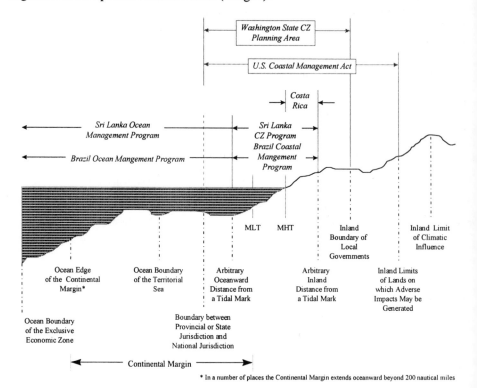

Figure 6.1. Existing and potential boundaries of the coastal areas. Re: late 1980s. Adapted from Sorensen and McCreary (1990: 8).

Economic Criteria
Looking at economic facilities, literature on coastal planning has regarded the land-ward limit of agricultural or rural coastal areas as relevant. Seawards, the outer limit of seaport installations, such as breakwaters and similar man-made structures, were considered as pertinent to management purposes.

Administrative and Jurisdictional Criteria
These have been more attractive than the economic ones. *Landwards*, the most limit of the coastal area was widely identified in the administrative areas, such as those of coastal municipalities, provinces, counties, and districts. *Seawards*, the limit of the internal waters, the outer limit of the outer continental shelf in the United States (3 nm from the baselines), and the outer limit of the territorial sea were regarded as relevant to the drawing of the maritime boundaries of the coastal area.

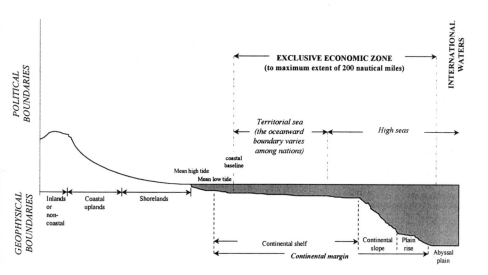

Figure 6.2. Boundaries pertaining to the delimitation of coastal areas: the physical and jurisdictional contexts. Adapted from Chua (1992: 32).

Leaving out the irrational arbitrary criteria, the conventional approach clearly concentrated on two kinds of limit.

First, approaches were attracted by limits drawn by the legal and jurisdictional frameworks. This field was interesting for two reasons. On the one hand, the limits of the coastal area, where coinciding with administrative boundaries (landwards) and jurisdictional boundaries (seawards), may objectively be defined and easily traced because they coincide with boundaries which pre-existed up to the adoption of the coastal management programme. On the other hand, these limits are concerned with the geographical stewardship of local authorities, i.e. a well defined decision-making system.

Secondly, coastal managers and planners have also been inclined to refer the limits of the coastal area to the physical features of the land and marine environments. This occurred specially where geomorphologists were involved in management programmes. In those cases the visibility of nature and the ease of assessing geomorphological and hydrological features, on which to base the limits, were the leading factors. As can be realised, Nature in that historical context was perceived in a limited sense, i.e. as consisting only of the abiotic components of the ecosystem.

The Role of Expanding Coastal Man-Made Structures and Uses
Looking at coastal planning the establishment of landward and seaward limits of coastal areas during the 1970s and 1980s has evolved because of the influence generated by significant growing economic and social changes. Unprecedented human pres-

sure on the coastal belts increased and the coastal systems expanded remarkably land-wards giving shape to vast coastal areas, including conurbations, metropolises and megalopolises. In the late 1960s and 1970s oil and gas offshore installations started expanding in many regional seas—from the south-eastern Asia to the Gulf of Mexico and the North Sea—and marked the tendency to widen coastal organisations seawards. There are some historical significance as, at that time, the *ekistics theory*, presented by the Athens Center of Ekistics, proposed the concept of *marine ecumenopolis* (Stewart 1970) to highlight the expansion of littoral settlements and growth of coastal populations. Meanwhile, another process, rich in cultural sense, came to the fore and diffused: the proclivity to establish protected coastal areas, such as reserves and parks. Their establishment would have induced coastal managers and planners to take endangered plant and animal species into consideration and thereby put the biotic components of the ecosystem in the agenda of their interests.

The evolution of criteria to delimit the coastal area has been associated with the evolution of management *raison d'être*, namely, the justification of the building up of coastal management programmes. Two trends have acquired increasing importance. On the one hand, there was the need to develop coastal economies by implementing such sectors as tourism, maritime transportation and seaports, and industry which had promising prospects. As a result, the delimitation of the coastal area was influenced by the need to focus on those land and marine environments which were considered useful for the development of the main sectors. On the other hand, there was the need to master physical processes, such as coastal erosion and lowland flooding. As human pressure on coastal belts grew, the risk of damage caused by these processes, including natural disasters was perceived as the more important socially. The key words marking these two trends were, respectively, "development" and "environment".

6.3. The Changing Design of the Coastal Area

As a consequence of Agenda 21 and subsequent guidelines on coastal management from inter-governmental organisations (FAO, IMO, UNEP, the Word Bank, and others), the political and scientific backgrounds to which the delimitation of the coastal area were referred have radically changed. The pursuit of sustainable development has led to focusing on the coastal ecosystem as the context in which a coastal management programme could be efficiently referred. In this respect, the optimum management would occur where the coastal area coincided with the geographical extent of an eco-system or set of contiguous ecosystems. Thereby, the coastal managers and planners are inclined to shape measures and define criteria guaranteeing the ecosystem's integrity and therefore protecting the basic resource for coastal development. This approach, where diffusely adopted, could lead to a remarkable improvement facing the conventional approach, tailored only or mainly to the abiotic niche of the ecosystem. The physical and chemical processes would be left in the background and the trophic webs regarded as the key subject for management.

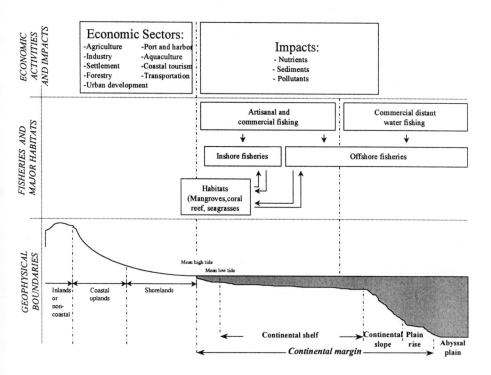

Figure 6.3. Boundaries pertaining to the delimitation of coastal areas: the physical and economic contexts. Adapted from Chua (1992: 32).

This innovative trend probably requires one or more generations before it wholly solidifies since assuming the extent of the coastal ecosystem as the geographical context for coastal management programmes is a very binding task. There are many reasons. For example, coastal management is clearly effective where the geographical coverage of the ICM programme coincides with the geographical stewardship of the local decision-making system. As a result, administrative areas and jurisdictional zones must continue to be assumed as first level factors. To combine these two spatial dimensions—i.e. that identified with ecological criteria and that provided by the administrative and jurisdictional frameworks—leads to stating that the delimitation of the coastal area could respond to this principle: *to minimise the spatial discrepancy between the geographical operational areas of the local leading decision-making system and the extent of the coastal ecosystem, or the extent of set of contiguous ecosystems.* This principle could be regarded as the equation of the *optimum delimitation of the coastal area.*

In this respect, reasoning shifts to two groups of limits respectively referred to social and juridical factors.

Landwards, first attention may concentrate on the limits of those land and brackish ecosystems where human settlements, other man-made structures and resource use extend. In many cases the landward limits of these ecosystems coincide with the coastal ecotone, namely, a typical ecosystem endowed with transitional features between marine and terrestrial ecosystems. As regards the Mediterranean, this kind of ecosystem can be found in the deltas of the Po and Nile, or in coastal areas characterised by the Mediterranean bush.

After the ecosystem, or the set of ecosystems, have been identified and delimited their geographical coverage may be related to the existing framework of administrative areas, such as coastal municipalities. The aim is to evaluate whether and how the coastal area may be delimited in such a way as to minimise the discrepancy between the ecosystem-based and administrative geographical extents. It may happen that a given administrative area, e.g. a coastal *département* (France), coincides with an ecosystem, or a set of contiguous ecosystems. Where it occurs, the optimum geographical coverage arises because optimum ICM can be pursued. More frequently, a group of contiguous administrative areas, e.g. contiguous municipalities, determines a geographical coverage not coinciding with the extent of the ecosystem, or a group of contiguous ecosystems. In these cases, coastal managers and planners should make efforts to encourage the relevant local decision-making centres to co-operate in order to ensure that, despite being included in different administrative areas, the ecosystem is managed efficiently.

Seawards, the framework of possible boundaries of the coastal area is more complicated than landwards. This is self-evident when the physical and jurisdictional lines, presented and discussed in Chapters 3 and 4, are borne in mind. Moving from those *mises au point*, three sets of lines are worth focusing on: (i) the land-sea interface lines; (ii) the water column depth lines; and (iii) the coastal distance lines.

6.4.   The Limits of the Coastal Area

6.4.1. THE LAND-SEA INTERFACE LIMITS

The coastlines and baselines are the main reference lines for coastal area management, the former being concerned with the physical context and the latter with the legal and jurisdictional frameworks. During the pre-take off and take off of coastal management (Table 1.2) these lines were frequently regarded as the seaward limits of the coastal area, since it was mostly consisting of emerged spaces. Now, instead, these lines are usually regarded as the landward limits of the marine part of the coastal area, since the management programmes are involved with increasingly extended marine spaces.

The Coastline
This physical line is an ambiguous reality because there are actually not only one but a group of different lines marking the land-sea interface. As can be seen from Figure 3.2, the backshore and foreshore may in fact be considered the transitional belt be-

tween the marine and land environment within which the coastline may be traced. Nevertheless, tracing this line on a map is a very onerous task because both the back-shore and foreshore are subject to uninterrupted changes occurring in to short-, mid- and long-term trends. Yet, the coastline plays a crucial role marking the transition between the terrestrial and marine ecosystems where fragile ecotones, involving the brackish, extend.

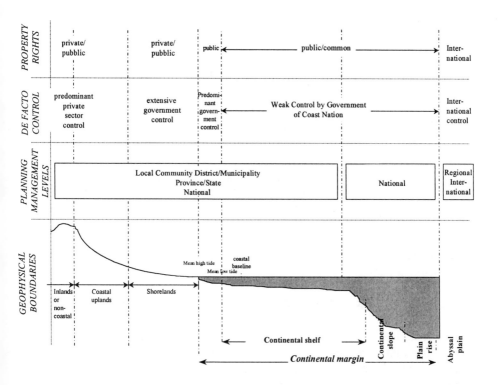

Figure 6.4. Boundaries pertaining to the delimitation of coastal areas: the physical, administrative and legal contexts. Adapted from Chua (1992: 32).

The Baseline

As has been seen in Chapter 4.2, the baseline only coincides with the coastline where the latter is straight, and not indented or fringed. Where this occurs, it is called "normal baseline"; where it does not, it consists of segments linking the outermost points of the coast, e.g. those delimiting bays and gulfs, and is called "straight base-line". Both normal and straight baselines have a role distinct from the coastlines since they separate the internal waters (landwards) subject to national ownership from ma-rine waters (seawards) subject to national jurisdiction. As a result, the coastline has a dual role for management. Where it does not coincide with the baseline, i.e. where straight baselines were established, it is only relevant in assessing the coastal ecosystem. Where the opposite occurs, i.e., where normal baselines were established, it is also relevant in assessing two different legal and jurisdictional milieus influencing coastal management.

TABLE 6.2.   The UN conferences on the Law of the Sea

Conferences	Main outcomes
First Conference;	Convention on the territorial sea and the contiguous zone;
1958	Convention on the high seas;
	Convention on fishing and conservation of the living resources of the high seas;
	Convention on the continental shelf.
Second Conference	No specific outcomes.
1960	
Third Conference	United Nations Convention on the Law of the Sea providing a comprehensive approach to the national maritime jurisdictional zones,
1973–1982.	the international waters and international co-operation on ocean protection.
Adoption: April 30, 1982.	
Entering into force: November 16, 1994.	

6.4.2. THE WATER DEPTH LINES

Water depth lines, which may be taken into account to delimit the coastal area, have a potentially wide range and are concerned with both the physical context of the continental margin and the national maritime jurisdictional zones. They are expected to become more and more relevant to coastal management.

The 200 m Isobath
This isobath has twofold relevance because it is concerned with the extent in depth of the coastal marine ecosystem and with that of the jurisdiction of coastal and archipelagic states on the seabed and subsoil.

Regarding the ecosystem, it approximately marks the lower limit of the neritic zone, where two kinds of ecosystems extend, brackish landwards and shallow marine waters seawards. Benefiting from the largest amount of solar energy, the ecosystems of the neritic province (see Chapter 3.3) have the most biomass of the ecosystems existing in waters overlying the continental margin. The 200 m isobath has acquired such increasing importance as fisheries and aquaculture have developed, to the point of being regarded as the key reference line for management of living resources.

As far as the state's jurisdiction is concerned, the 200 m isobath has served as the reference line for coastal and archipelagic states to explore and exploit oil and gas fields. This has been due to the 1958 Convention on the continental shelf (Table 6.2), which had helped open the way to developing coastal ocean mining. Since then, the jurisdiction on seabed and subsoil has been included in the agenda of national economic programmes, and has led coastal managers and planners to focusing on the 200 m isobath for delimiting the coastal area.

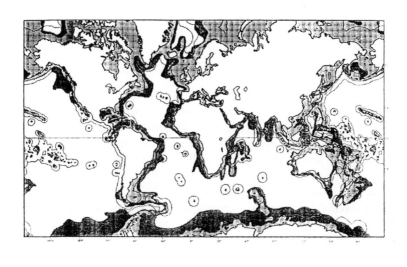

CONTINENTAL PLATE   CONTINENTAL MARGIN   ——————— 200 nmi

Figure 6.5. The extent of the world continental margin and the 200 nm isodistance line from the baselines.

The Outer Edge of the Continental Margin

This line, extending at various depths (up to 4,000–5,000 m) according to the existing tectonic plates, is a belt actually characterised by great quantities of sediments and marks the transition between the continental marine environment and the deep-ocean. As a consequence, it gives shape to the transition between the coastal ocean and deep-ocean ecosystems. Furthermore, it is important in relation to the jurisdictional framework and exploration of mineral resources. This is due to the 1982 LOS Convention, which enables coastal and archipelagic states to explore and exploit mineral resources—currently, only oil and gas fields are concerned—up to the outer edge of the margin provided that it does not extend beyond the 200 nm line from the baselines. The outer edge of the continental margin is therefore only relevant to coastal management where it is located landwards from the 200 nm isodistance line. As already noted (Chapter 4.5), according to the law of the sea, the outer edge of the continental margin is legally fixed on the 2,500 isobath.

6.4.3. THE DISTANCE LINES

Where attention shifts to lines based on distance, expressed by isodistance lines in maps, three lines strongly influencing coastal management come to the fore: the 12 nm, 24 nm, and 200 nm equidistant lines from the baselines. Evidently, they have become increasingly important in delimiting the geographical seaward coverage of ICM programmes.

The 12 nm Line
This line, which is 12 nm from the baselines, is particularly significant to coastal management as—in accordance with LOS Convention—it marks the outer limit of the territorial sea. Since the take off stage of coastal management, states have tended more and more to adopt the boundary of the territorial sea for management purposes. This is widely motivated since, within its territorial waters, the coastal and archipelagic states are fully capable to manage all the biotic and abiotic components of the coastal ecosystem (Chapter 4.3). In addition, the territorial waters frequently embrace most of the neritic zone, rich in biomass and living resources.

The 24 nm Line
This line marks the outer limit of the contiguous zone (Chapter 4.4). Although it only played a minor role in coastal management in the past, it has recently become important for two reasons. First, coastal and archipelagic states have taken action to keep these waters under control against customs, sanitary and fiscal infringements. Secondly, according to the juridical doctrine, the 24 nm isodistance line also delimits the so-called archaeological zone, therefore giving the coastal and archipelagic state jurisdiction to protect the cultural submarine heritage. This jurisdiction is expected to increase rapidly because of the enormous progress in techniques used to identify and explore undersea settlements and man-made structures, and to raise wrecks.

The 200 nm Line
After the 1982 LOS Convention this line was included in the agenda of coastal managers as a reference basis to trace the outer limit of:
a) the continental shelf, profoundly innovating the 1958 Convention approach (Chapter 4.5) which was based on depth (200 m isobath);
b) the exclusive economic zone (EEZ); therefore opening the way to the claim to
c) the 200 nm wide exclusive fishery zones (EFZs).

As regards the continental shelf, the role of the 200 nm isodistance line was that of prolonging the capability of the coastal state to exploit oil and gas fields beyond the limit established by the 1958 Convention. However, where EEZs were not claimed, the (jurisdictional) continental shelves only partially concern coastal margins because they enable the coastal state to exploit only the mineral seabed and subsoil resources. The sea surface and water column have not been included in the state's jurisdiction. Instead, where the 200 nm line marks the outer limit of the EEZ, it influences coastal management integrally because the coastal and archipelagic states have full management capability (Chapter 4.6). It may be recalled that Agenda 21, Paragraph 17.3, explicitly considers the EEZ as an important geographical coverage of the coastal area.

6.5.   The Search for Rationale and Effectiveness

As can be seen from the above discussion, a wide range of criteria can be adopted to delimit the geographical coverage of ICM programmes. Nevertheless, choosing among them is not by chance, at least when Agenda 21 is assumed as the basis for the approach. In order to choice from different criteria, the concept of *optimum geographical extent* of the coastal area needs to be looked into before options are adopted. As a preliminary approach, it could be stated that *optimum geographical coverage occurs where the coastal decision-making system is enabled to build up management as consistent as possible with the integration principle.*

The integration-inspired approach moves from the prospect of minimising the discrepancy between the geographical extents of the ecosystem, or a group of contiguous coastal ecosystems, and that of a national maritime jurisdictional zone. The ecosystem is usually considered moving from its abiotic niche, implying that geomorphologic, geological, hydrological and climatic features and processes are taken into account. However, contrary to what has occurred in previous historical phases of coastal management, these features and processes should not be considered *per se*, but to focus on the trophic webs, i.e. the biotic component of the ecosystem. To be explicit, the concept of the abiotic niche should only be functional to that of the biotic organisation.

In this respect there is no doubt that, seawards, attention is to be concentrated on the continental margin because it is the niche of the set of marine coastal ecosystems. It follows that the jurisdictional zones with the capability of managing the continental margin as a whole, or at least the continental shelf, have become all the more important after adopting Agenda 21. So, it is self evident that *optimum coastal management exists where the EEZ was claimed and the width of the (physical)continental shelf, or continental margin, is approximately 200 nm.* As a matter of fact, in this case a wide abiotic niche, including the coastal ecosystem or set of contiguous coastal ecosystems, is covered by a jurisdictional zone enabling the coastal state to develop full management, from sea surface to seabed and subsoil.

TABLE 6.3. The sea lines relevant to coastal management

Lines		Reference to	
		Ecosystem	National maritime jurisdiction
Isodistance lines from the baseline			
12 nm			Territorial sea
24 nm			Contiguous zone
200 nm			Continental shelf; EEZ; EFZ
350 nm			Continental shelf
Isodistance lines from the 2,500 m isobath			
100 nm			Continental shelf
Isobaths			
200 m		Neritic zone	Continental shelf

Unfortunately, as Figure 6.5 shows, this coincidence is rarely found, and it is in fact an extreme case. Most of the ocean world is characterised by physical continental shelves, or continental margins, extended seawards or landwards from the 200 nm isodistance line marking the outer limit of the EEZ. As a consequence, coastal managers have to conduct fairly difficult exercises, where the extents of coastal ecosystems' abiotic niches are compared with that of the EEZ with the aim of designing rational seaward boundaries of the coastal area.

Where the EEZ was not claimed or agreed upon, and the jurisdictional continental shelf exists, the issue is even more complicated. First, it is an extreme case that the jurisdictional continental shelf coincides with the physical continental margin or shelf. Secondly, where this occurs, the jurisdictional continental shelf enables the coastal and archipelagic state to manage only a part of the ecosystem, consisting of the seabed and subsoil. Therefore, the most important component of the ecosystem, consisting of the trophic webs living in the water column, is excluded.

Therefore, to respond to the needs for ICM, efforts should be made to claim, or agree upon, jurisdictional zones covering the geographical extent of an individual ecosystem, or set of contiguous ecosystems. The framework of the national maritime jurisdictional zones existing in the individual coastal sea should be considered from this point of view and should be arranged to optimise interaction between the juridical and jurisdictional factors, on the one hand, and the ecological principle, on the other. Nevertheless, this goal is difficult to achieve for many reasons. Firstly, the decision-making centres should perceive that the economic interests are better protected when the ecosystem is managed in a holistic way, and that the framework of lines delimiting the national jurisdictional framework should be viewed innovately. Secondly, social perception is also needed. Coastal communities should become aware that the maritime jurisdictional framework plays a vital role in coastal management and that the geographical coverage of ICM programmes is not only a technical concern, but also an issue which, depending on how it is approached, can have important consequences on social life, including that of future generations.

This reasoning leads us to understand how complicated the geographical delimitation of the coastal area is, both landwards and seawards, and how it requires great attention. There are many possible approaches leading to numerous results due to national policy and subnational strategy, ending with social perception. Facing the complication due to the existence of this web of factors, scientific literature may provide only basic models useful for designing approaches area by area. Complication is concerned with both the land and marine parts of the coastal area but, as has been noted, from the late 1980s the issues concerned with the maritime milieu have been acquiring special importance for two reasons. First, in this space, a wide spectrum of geomorphologic, hydrological, ecological and jurisdictional zones exists, each of them deeply influencing approaches from coastal managers. Secondly, an increasing number of coastal communities have perceived the coastal sea as the main resource for their development.

To deal with this changing framework, the design for the geographical coverage of the marine coastal area has to take into account some of the following fundamental criteria, based on distance and depth.

6.5.1. LANDWARDS

The administrative or political boundaries might be the reference lines of the first approach. This would enable managers and planners to refer to the existing geographical web of administrative areas where an existing decision-making system is able to exert authority, and thereby ensure the ICM programme is adopted and implemented. The geographical extent of administrative areas needs to be analysed with the aim of designing a geographical coverage of the ICM programme constituted by administrative areas and, contextually, encompassing a coastal ecosystem, or set of coastal ecosystems. This enables the primary principle of ICM to be met which consists in protecting the ecosystem's integrity.

This process requires the ecosystem to be delimited. In this respect, attention may initially be paid to its abiotic niche. For example, as a preliminary step, the watershed of rivers flowing into marine coastal waters could be regarded as relevant. The alternative approach would consist of focusing directly on the extent of trophic web communities, i.e. the flora and fauna habitats.

The final goal of this exercise is to minimise the geographical gap between administrative and ecological areas.

6.5.2. SEAWARDS

The spatial delimitation of the coastal area might refer to this range of jurisdictional lines.

The 200 nm Isodistance Line

Where possible, the best approach might consist in setting up ICM programmes tailored to the 200 nm width from the coast. In practice, the seaward limit of the geographical coverage should be the 200 nm isodistance line from the baseline. This presupposes that the EEZ, or the EFZ up to 200 nm, is established. If the EEZ operates, full management, from sea surface to subsoil, can be conducted; if only the EFZ is established, management may be concerned only with the water column living resources.

The 12 nm and Other Isodistance Lines
If an EEZ was not established and the jurisdictional continental shelf was claimed, the maritime extent of the coastal area might be based on one of the following three limits:
a) the 12 nm isodistance line from the baselines, coinciding with the outer limit of the territorial sea and delimiting that part of the marine area where any component of the ecosystem can be managed;
b) the outer limit of the jurisdictional continental shelf, circumscribing that part of the area where only the seabed and subsoil can be managed;
c) the outer limit of the EFZ, if not extending up to the 200 nm isodistance line, which delimits that part of the coastal area where only living resources can be managed.

Opposite or Adjacent Coasts
Where opposite or adjacent coastal states exist, the coastal manager has to consider the boundary which they agreed upon to delimit their own jurisdictional continental shelves. In this case the agreed median line is the outer limit of the coastal area, and the limits of the opposite or adjacent coastal areas coincide with the same line, as happens in the land boundaries. This occurs in many semienclosed and enclosed seas, where the states have opposite or adjacent coasts less than 400 nm away. The Mediterranean belongs to this group of seas and the continental shelves may be only agreed in this basin. As a result, the Mediterranean coastal areas are expected to coincide mostly with the median line between opposite and adjacent states. Until now, only a limited part of the Mediterranean coastal regions have been endowed with agreed jurisdictional continental shelves.

The Neritic Zone
Whatever the existing framework of the national maritime jurisdictional zones is, management programmes should be tailored to the neritic zone, so the 200 m depth line would be the basic reference. The geographical extent delimited by the 200 m isobath differs according to the areas as a result of the different gradient of the continental shelf and slope, over which the neritic zone extends. After delimiting the neritic zone, the web of jurisdictional zones which it is included should be identified. According to the local conditions, the neritic zone could be seen as encompassed only by the territorial sea or as extending seawards from the outer limit of this jurisdictional zone. In the latter case, it could partly be included in the jurisdictional continental shelf, the EEZ or EFZ. As can be seen, quite a wide range of situations may occur depending on the geomorphologic features of the continental margin and the existing framework of the jurisdictional zones. As a consequence of this analysis, the coastal manager knows the extent to which the neritic zone can be managed. If this

zone is included only in the territorial sea or within the EEZ, full management can be conducted. Where included in the territorial sea and the continental shelf, it consists of two parts: landwards from the outer limit of the territorial sea, full management is practicable; seawards from this limit, only partial management is possible.

To conclude on how the geographical delimitation of coastal areas may be approached, it is worth noting that a useful technical tool may be provided by coastal GIS. It makes it possible for the set of ecosystem-referred, administrative and jurisdictional lines to be represented and correlated with lines pertaining to other features of the coastal area, such as urban cover and land uses. Besides that, using simulation techniques, the possible extent of the coastal area for management purposes may be traced and the optimum geographical coverage of the ICM programme may be identified.

6.6.   The Mediterranean as a Case Study

The approach presented in this Chapter may be the regarded as a basis for two distinct operations. On the one hand, the concepts concerning the delimitation of the coastal area may be discussed in order to design the best methodological approach to the geographical coverage of the individual ICM programme. On the other hand, the delimitation as proposed in the previous section may be applied to an individual coastal area to evaluate how useful and effective it is. The Mediterranean sea is a good example of this because, when considering its coastal management evolution (Chapter 1.8), three focal questions arise: (i) whether and how the guidelines for ICM provided by UNEP (1995) efficiently deal with the delimitation of the coastal area; (ii) how the coastal areas, which have benefited from coastal management programmes carried out in the MAP framework, have been delimited; (iii) what issues are expected to arise where the approach proposed in this Chapter (Section 5) were applied.

6.6.1. CRITERIA AND GUIDELINES BY UNEP

The UNEP guidelines (UNEP 1995: 51–2) provide a definition of the coastal area and distinguish it from the coastal zone. This approach was presented in detail in Chapter 1.5. Considering the approach of USA literature, such as by Sorensen and McCreary (1990), UNEP states that a "*Coastal area* is a notion which (...) indicates that there is a national or sub-national recognition that a distinct transitional environment exists between the ocean and terrestrial domains". This concept differs from that of coastal zone, "the borders of which require a less strict definition". Nevertheless, Figure 1.2 presenting this approach provides a very wide geographical coverage of the coastal zone, extending between the outer limit of the coastal waters to the watershed, so including all the drainage basins. It is therefore difficult to imagine that the coastal area is a wider geographical extent than the coastal zone.

From these considerations, attention shifts to another part of the UNEP guidelines (1995: 25), where it is recommended that the materials relevant to the adoption of individual ICM programmes provide "precise definition of the coastal area, i.e. the area boundaries (for example, in the case of areas covering the totality of a national coastline it is desirable to include the relevant drainage basins and the national maritime exclusive economic zone; in the case of small island states, the inclusion of the whole national territory is advantageous)". This approach leads to believing that each decision-making system—the government in the case of the national master plan on ICM; a sub-national authority such as a *Regione* in Italy—feels free to adopt any kind of delimitation criterion. As regards the Mediterranean, it is unlikely for reasons discussed in Chapter 4.8, that EEZs will be established, although they could contribute to sustainable development of marine spaces. This approach gives the impression that UNEP was not keen to look too closely at the delimitation issue of the coastal area and, more importantly, that the relevant guidelines were not designed as a follow-up of a critical view of literature and existing practical approaches facing the principles from Agenda 21. This lack of attention may also be confirmed by the database that UNEP (1995: 37-9) suggested for ICAM development, which does not include data pertaining to the ecosystems, nor the administrative and jurisdictional frameworks.

6.6.2. THE MAP PROGRAMMES ON ICM

As seen in Chapter 1.8, the MAP experience has been carried out through the Coastal Area Management Programmes (CAMPs). Their main fields were: (i) implementation of legal instruments; (ii) resource evaluation, protection and management; (iii) activities (evaluation and trends); (iv) natural hazards and phenomena; (v) planning and management tools; (vi) development-environment scenarios; (vii) integrated planning and management (UNEP(OCA)/MED IG.5/Inf.3, draft 1995: 85). Evidently, the subject areas relevant to the delimitation of the coastal area cannot be found. The CAMP development has gone through some phases presented in Table 1.8. None of these perceived the coastal area delimitation as an issue to be tackled with well defined methodologies. For example, the two major programmes relating to the Bays of Izmir (see especially MAP/PAP 1994) and Kastela (University of Split 1994), are concerned:

a) *landwards*, with the metropolitan areas, as they are defined by approaches of local authorities;

b) *seawards*, with the geographical extent of the bay.

A critical evaluation of the role of delimitation facing the effectiveness of the programme cannot be found in these materials. However, 1995 saw the third generation of CAMPs designed to pursue ICM. Based on stimuli provided by both Agenda 21 and literature, the delimitation issue is expected to be dealt with in this newly-convened generations of programmes.

6.6.3. POSSIBLE APPROACHES

In the Mediterranean, the *landward* delimitation of the coastal area for management purposes should integrate three criteria:

a) the extent of the land coastal ecosystems, such as the Mediterranean bush or forest;
b) the administrative boundaries within which the decision-making systems involved in coastal management extend their stewardship;
c) the extent of land settlements, facilities and natural resource uses.

The situations vary too much according to the individual and so numerous case studies can be found.

The *seaward* delimitation should integrate:
a) the extent of the marine coastal ecosystem, which is to be basically referred to the neritic zone and the 200 m isobath;
b) the extent of the widest national jurisdictional zone;
c) the extent of the physical continental shelf over which the economic interest of states (oil and gas fields) more extends.

In this part of the coastal area the delimitation issue is more intriguing and has not been dealt with yet. Besides, the situations do not vary so much they do landwards. As a result, some typical cases may be identified.

The Central and Northern Adriatic Sea
Most of this water column overlies the continental shelf and is no more than 200 m deep. The jurisdictional continental shelves were established, so the Adriatic waters were divided into two parts with the median line (Figure 4.5). The coastal area may extend up to this line but the coastal states within their jurisdictional continental shelves cannot manage the sea surface and water column, so only partial management can be carried out between the 12 nm line (outer limit of the territorial sea) and the Adriatic median line. Where states facing this sea could agree on establishing the EEZs following the boundary of the existing continental shelves, these maritime spaces could benefit from the optimum coastal area extent. However, the geopolitical context, influenced by the collapse of the former Yugoslavia, implies that this is highly unlikely.

Aegean Sea, Turkey
The 200 m isobath in the Aegean coastal areas extends beyond the outer limit of the territorial sea, which was established 6 nm from the baselines. As a result, the coastal area cannot be delimited so as to include the whole extent of the marine coastal eco-system. Therefore, Turkey is probably in the worst position in the Mediterranean with regard to operating ICM.

Eastern Mediterranean, Alexandria Area
The 200 m isobath is included within the territorial sea (12 nm). No jurisdictional continental shelf was established. As a result, the marine coastal ecosystem could be managed efficiently if ICM programmes embracing the territorial sea were under-taken.

This first approach to the Mediterranean context, if associated with the general discus-sion provided in the previous Sections, leads to some basic conclusions:

a) the geographical delimitation of the coastal area is an unavoidable issue whose solutions deeply influence the effectiveness of the ICM programme;

b) this issue needs to be dealt with using multi-disciplinary approaches at least involving national law, administrative setting (land), law of the sea and, especially, ecology;

c) each coastal area needs its own approach but this does not imply leaving out the national and multi-national contexts of reference;

d) rational criteria, as consistent as possible with the guidelines and spirit of Agenda 21, need to be adopted;

e) a range of possible limits of the coastal area may be provided, each defined according to the sustainable development concept;

f) the rational geographical extent of the coastal area needs to be evaluated while the ICM programme is implemented and, if necessary, adjusted to optimise management effectiveness.

CHAPTER 7

THE COASTAL USE STRUCTURE

This Chapter will introduce the concept of coastal use structure to investigate and represent the kinds of natural, human and social resource exploitation on which the socio-economic organisation of the coastal area is based. A methodological approach to this subject will be proposed consistently with the aim of building up ICM programmes. The following questions will be discussed:

Questions	Figures	Tables
How can the holistic view of the coastal uses be constructed?		7.2, 7.3
Which concepts of coastal use and coastal use structure are useful for ICM?	7.1	7.1
How may the coastal use structure be assessed?	7.2	
How has literature dealt with the subject?		7.2, 7.3
How may the coastal uses be classified?	7.3	7.4, 7.5
How may relationships between coastal uses be assessed and represented?	7.3	
How may impacts from coastal uses on the coastal ecosystem be approached?		7.4
Which options for ICM may be highlighted in assessing the coastal use structure?	7.4	
What is the result?	7.5	

7.1. The Conceptual Framework

In Chapters 1 and 3 the coastal area was presented as a complex, bi-modular system integrating one or more ecosystems and the coastal use structure. Now, attention focuses on the coastal use structure which may be assumed as consisting of:
a) a set of uses practised in the coastal area, i.e. *the coastal use framework*;
b) a set of relationships between uses, i.e. the *horizontal relationship web* of the coastal use structure; and
c) a set of relationships between coastal uses and the coastal ecosystem; i.e. the *vertical relationship web* of the coastal use structure.

The coastal use structure does not coincide with the coastal system because the coastal system is thought of as the coastal structure *plus* its organisation, *plus* the goals towards which the coastal structure moves because of its organisation. The assessment of the coastal use structure is indeed essential to understand and design the coastal system. Consequently, it plays a vital role in the undertaking and carrying out of ICM programmes. To respond to the needs to set up integrated approaches to coastal management the most fruitful path is to consider, in sequence: (i) the coastal

143

use framework, (ii) the horizontal relationship web, and (iii) the vertical relationship web.

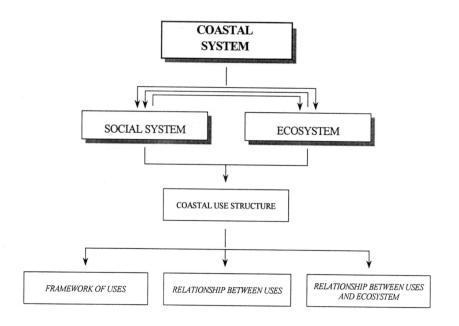

Figure 7.1. The conceptual approach to the coastal use structure.

7.2.   The Coastal Use Concept

The coastal use framework is the set of coastal uses regarded *per se*, i.e. for the moment leaving out the relationships between uses. The building up of the coastal use framework, which is essentially a taxonomic subject, implies that the concept of coastal use is presented and discussed first. This concept may be designed taking into account and rationalising a wide spectrum of contributions from literature on coastal management concerned with the resource use, facilities and activities. Following Sorensen and McCreary (1990: 12-3), the assessment of coastal areas may be set up considering (i) the coastal resources, (ii) the coastal uses, and (iii) the coastal ecosystem. A coastal resource is defined "as a natural, often renewable commodity, the existence of which depends on the coast or on the value attributed to the circumstance that it is located within the coastal zone. Sometimes constructed features such a scenic coastal village are included in the definition of coastal resource". Moving from this concept, the "coastal use refers to the utilisation of coastal resources for economic, aesthetic, recreational, scientific or educational purposes". This is a useful *mise au point* of the conceptual bases which have been developed by leading coastal management programmes. If this definition is adopted, the concept of use clearly presupposes two other concepts: resource, and goal to be achieved with the resource use.

From the epistemological point of view this reasoning evokes the rational mechanistic paradigm (Le Moigne 1977–1994: 47–8), according to which reality consists of structures generating functions. In fact, the definition by Sorensen and McCreary considers the coastal use as a *function* of the resource and a human exploitation *tool*. This can be presented in a Cartesian diagram in which:

- the X axis pertains to the coastal resource;
- the Y axis pertains to the tool with which the coastal resource is exploited.

At the convergence of the X and Y axes, U points represent the coastal uses. This means that coastal use is a characteristic *relational concept* because it can be defined only with reference to other concepts, in our case those of resource and exploitation tool. Nevertheless, the image of the Cartesian diagram should not deceive us. In spite of this simple representation, the concept of use is ambiguous since it hides another idea, that of goal. In fact, each resource use is developed to meet an individual need. For example, fish farming (coastal use) is at the cross point of living resource (X axis) and biotechnological tools (Y axis), and it is practised to meet the social demand for seafood (social goal). As a result, the diagram should really include *goals* instead of uses. Here the word "use" is kept so as not to abandon current terminology.

This phase of reasoning is crucial since, when use is considered as mirroring a goal, attention clearly has to focus first on the goal system of coastal area management. According to how the goal system is determined, the approach to coastal management and consequently methods vary in constructing the use framework. Doubtless to say that each coastal community pursues its own goals. According to the general system theory, a system—in our case the coastal system—can move towards two categories of goals:

a) an individual goal, or a set of goals, explicitly determined—in this case the system is characterised by *regulated teleology*;

b) an individual goal, or a set of goals, not explicitly determined—in this case the system is characterised by *wild teleology*.

Following this approach, coastal area management, where it is conducted through the building up of coastal management programmes, can be considered as *the shift from a wild coastal system to a regulated system* (Vallega 1996: 99–100).

Reasoning leads us to take into account that the coastal use framework varies from area to area in relation to the goals pursued through coastal management. So the best way to present the coastal use framework is to focus on the goals pursued in the historical phase under analysis. Goals are determined by both the local decision-making centres and decision-makers acting at other levels. In a general sense the determination of goals depends on the association of inputs from the global, regional (multinational), national and subnational scales (Table 7.1).[34]

For example, the coastal use framework in the Mediterranean Sea is to be related to two distinct levels of determining goals according to whether the coastal area is located within or outside the European Union (EU) space. Where coastal and archipelagic regions belong to the EU, the goals determined by local decision-making centres should actually be consistent with UNCED (Agenda 21, Chapter 17) but also with the guidelines provided by the Mediterranean Action Plan (MAP), the regulations from the EU and national policy. Where the state is not an EU member, the framework is less complicated because there is not need to conform to EU policy.

TABLE 7.1. Determination sources of coastal management goals

Levels of determination	Examples of goals
Local decision-making systems, such as *départments* in France, *Regioni* in Italy, or the States of a Federal system, such as in the USA.	Waterfront revitalisation, carried out by the local municipality and port authority.
Governmental organisations.	Following the guidelines from the Ministry of energy, installation of coastal thermal plants using imported oil and serving the national market.
Inter-governmental organisations operating on the regional (multi-national) scale.	Establishment of specially protected areas as a response to the recommendation by the Special Protected Areas, a Regional Activity Centre of the Mediterranean Action Plan.
Inter-governmental organisations acting on the global scale.	As a result of the collaboration with the International Maritime Organisation, location and management of coastal waste disposal according to the prescriptions of MARPOL.

Once the determination levels of coastal management goals have been identified, the consistency of the individual coastal uses may be assessed and mentioned in the coastal use framework. For example, if trawl-net fishing is practised in an individual Mediterranean coastal area, it should be specified in the coastal use framework that this coastal use is prohibited by both those international and national regulations which were designed to protect the marine biotic communities. As a result, it is not consistent with ICM goals. As can be seen, when the coastal use framework is constructed, the individual coastal uses may be marked by their consistency with the international, multi-national and national regulations and the principle of integrated management. This is a preliminary step to evaluate the role of the uses existing in the coastal area where a ICM programme is undertaken. Hence, it provides the basis to re-structure the web of coastal uses on which the organisation of the coastal area is founded.

7.3.   The Classification Issue: Theory and Practice

To carry out this evaluation it is necessary to deal with the kind of resource each coastal use relates to, on the one hand, and the objective each tends to pursue, on the other. This approach has many operational consequences because the sustainable development principle requires optimising resource exploitation explicitly presenting the objective to be pursued by management. It implies coastal uses are classified in order to design coastal use frameworks for management programmes in a useful logical way. So the preliminary task is to look into the classification methods.

TABLE 7.2.   Coastal use frameworks in literature

Global Marine Interaction Model (GMIM)[1] 1983	Sorensen and McCreary [2] 1990	Pido and Chua [3] 1992	Coastal Use Framework [4] 1992
1. Navigation and communication	1. Fisheries	1. Agriculture	1. Seaports
2. Mineral and energy resources	2. Natural area protection systems	2. Fisheries and aquaculture	2. Shipping, carriers
3. Biological resources	3. Water supply	3. Infrastructure	3. Shipping, routes
4. Waste disposal and pollution	4. Recreation development	4. Mining	4. Shipping, navigation aids
5. Strategy and defence	5. Tourism development	5. Ports and harbours	5. Sea pipelines
6. Research	6. Port development	6. Industry	6. Cables
7. Recreation	7. Energy development	7. Tourism	7. Air transportation
8. Environment	8. Oil and toxic spill contingency planning	8. Urban development	8. Biological resources
9. Management	9. Industrial siting	9. Forestry	9. Hydrocarbons
	10. Agricultural development	10. Shipping	10. Metalliferous renewable resources
	11. Mariculture development		11. Renewable energy sources
			12. Defence
			13. Recreation
			14. Waterfront man-made structures
			15. Waste disposal
			16. Research
			17. Archaeology
			18. Environmental protection and preservation

[1]   Couper A.D. (ed.), *The Times Atlas of the Oceans*. Times Books, London, 1983, p. 208. This model is presented as a square matrix and assumes that the conservation and management of the natural coastal environment also belong to coastal zones.
[2]   Sorensen J.C. and McCreary S.T., *Institutional Arrangements for Managing Coastal Resources and Environments*. National Park Service, US Department of Interior, US Agency for International Development, Washington DC., 1990, pp. 41–2.
[3]   Pido M.D. and Chua T.-E., "A Framework for Rapid Appraisal of Coastal Environments", in: Chua T.-E. and Fallon Scura L. (eds) (1992) *Integrative Framework and Methods for Coastal Area Management*. Manila, International Center for Living Resources Management, 1992, pp. 144–7.
[4]   Vallega A., *Sea management. A theoretical approach*. Elsevier Applied Science, London, 1992, Chapter 7, Appendix D.

Classifying coastal uses is not a recent concern and literature began dealing with it in the early 1980s. A basic approach may be found in the *Global Marine Interaction Model* (Couper ed. 1983: 208) which had a main role in concentrating attention on matrix-based methodologies. At the moment, four classifications presented in Table 7.2 are worthy of consideration. It can be noted that they respond to different conceptual approaches.

The Global Marine Interaction Model (GMIM) does not differ from others just because it is concerned with both coastal and deep ocean uses but also because it focuses on (i) branches of the local economic organisation (navigation and communications, research, recreation and others), (ii) categories of natural resources (biological, mineral, energy resources), and (iii) waste disposal. The framework by Sorensen and McCreary (1990) was set up to show the issues (such as coastal erosion) and economic prospects (such as recreation and development) which, at that time, were perceived by decision-making centres and coastal managers as extremely important. As a result, most categories consist of (i) economic sectors (tourism, ports, agriculture and others), while the rest are concerned with (ii) resource exploitation (energy, water), and (iii) environmental protection (natural area protection systems). The framework by Pido and Chua (1992: 144-7) also includes activity sectors and natural resources. Vallega's framework (1992) includes (i) resources (biological, mineral, energy resources), (ii) branches of the economic organisation (seaports, shipping, air transportation), (iii) man-made structures, and (iv) environmental protection.

There is no doubt that the *a priori* and deduction-inspired frameworks provided by the literature, which mainly aims to construct bases for analysing case studies, needs to be related to significant classifications adopted by coastal management programmes, or by programme guidelines. In this respect three classifications are presented in Table 7.3. They reflect (i) the approach developed in the USA as a consequence of the Coastal Zone Management Act (CZMA, 1972) and following developments, (ii) an approach due to the efforts of south-eastern Asia decision-making systems, in their turn more or less influenced by the USA experience, and (iii) the approach recommended by UNEP.

Although comparison could be extended to other coastal management programmes, the above trio is useful for examining how the design as well as concept of the coastal use framework change according to the features of social and geopolitical contexts. The Lingayen Gulf programme is keen on the need to achieve sustainable development of biological resources *vis-à-vis* the increasing human pressure on the coastal area. The Hawaii programme reflects the need to diversify the existing economic organisation implementing tourism, recreational uses and landscape protection, as well as to encourage marine mineral (nodules) and energy (marine thermal gradient) exploitation. The UNEP approach would like to deal with the uniqueness of coastal areas subject to increasing human pressure, economically diversified regional organisation and maritime bulk transportation.

By comparing the coastal use frameworks presented in Table 7.2 and 7.3, a further deduction can be drawn on how the coastal area is perceived by managers. Most categories of uses included in the Global Marine Interaction Model are concerned with the coastal sea. This framework was designed by scientists who perceived the coastal area as a space mainly providing marine resources; hence a sea-oriented classification (seaport, navigation, offshore industry, etc.). On the contrary, the classification adopted by UNEP focuses on the exploitation of the land coastal area and does not give much room to sea-based uses. These different orientations basically depend on how the role of coastal resource exploitation is perceived.

TABLE 7.3. Coastal use frameworks according to coastal management programmes

Hawaii Ocean Resources Management Plan [1] 1991	Lingayen Gulf Coastal Area Management Plan[2] 1992	United Nations Environment Programme (UNEP) [3] 1995
1. Research	1. Fisheries	1. System of urban and rural centres
2. Recreation	2. Alternative livelihood for fishing families	2. Open spaces
3. Harbours	3. Aquaculture	3. Agricultural land
4. Fisheries	4. Environmental quality	4. Forestry
5. Marine ecosystem Protection	5. Critical habitats	5. Mining
6. Beaches and coastal erosion	6. Linked habitats	6. Industrial areas
7. Waste management		7. Residential areas
8. Aquaculture		8. Tourism and recreational areas
9. Energy		9. Sea uses
10. Marine minerals		10. Transport corridors and areas
		11. Other infrastructures

(1) Hawaii Ocean and Marine Resources Council, *Hawaii Ocean Resources Management Plan, Technical Supplement*. Hawaii Ocean and Marine Resources Council, Honolulu, 1991.
(2) National Economic Development Authority (Region 1, Philippines), *The Lingayen Gulf coastal area management plan*. International Center for Living Resources Management, Technical Reports, No. 32, Manila, 1992. See also: Scura L.F., Chua T.-E., Pido M.D. and Paw J.N. (1992) "Lessons for integrated Coastal Zone Management: The ASEAN Experience", in: Chua T.-E. and Scura F.L. (eds) (1992) *Integrative Framework and Methods for Coastal Area Management*. Manila, International Center for Living Resources Management, pp. 1–70.
(3) UNEP *Guidelines for Integrated Management of Coastal and Marine Areas - with Special Reference to the Mediterranean Basin*. UNEP Regional Seas Reports and Studies, No. 161. Split, PAP/RAC (MAP-UNEP), 1995. This approach was developed moving from the experience carried out by the PAP/RAC of MAP.

7.4.   The Coastal Use Framework

The above presentation of frameworks may show how the choice between the criteria to classify the coastal uses is methodologically intriguing. Shifting from one criterion to another the resulting framework changes thus influencing the whole design of the ICM programme. The lesson to draw is that the construction of frameworks requires paying great attention to the consequences of the individual methodological approach.

Where integrated management is pursued, the framework of coastal uses has to be constituted in such a way as to present the resource exploitation in the light of the three components of sustainable development, i.e. (i) integrity of the ecosystem, (ii) economic efficiency, and (iii) social equity including rights of future generations. Following Young (1992: Chapter 2), the integrity of the ecosystem is guaranteed when human actions capable of altering the ecosystem's organisation and influencing its evolution are avoided. Human influence may be evaluated using biological parameters, essentially productivity, biodiversity and resilience. Economic efficiency can be regarded in terms of both productive and allocative efficiency. The former principle "requires that waste is avoided and that production increases until marginal costs equal marginal revenues or benefits". The latter principle "requires that there is no alternative allocation which would make at least one person better off and no one worse off" (Young 1992: 27). Social equity is referred both to present communities and future generations. Equity within present communities, i.e. intra-generation equity, concerns the "just distribution of resources, rights and wealth among people and over time". Inter-generational equity is based on two principles: "each generation should maintain the resource base it inherits and leave the next generation with a per capita stock that is no less than it inherited; and the total stock of conditionally-renewable resources, resource diversity and assimilative capacity should be maintained through time" (Young 1992: 37–41).

From these statements it follows that ICM is referred to these hierarchical systems of goals:
- *first rank*: integrity of the ecosystem
- *second rank*: social equity
- *third rank*: economic efficiency.

In practice, (i) social and economic objectives can be optimised only if the ecosystem's integrity is guaranteed, (ii) social equity is the final end of policy, and (iii) economic efficiency is considered as a tool to protect the ecosystem and to reinforce social equity.

This reasoning leads us to design coastal use frameworks like that presented in Table 7.4. This framework is an *a priori* classification of coastal uses whose aim is to serve as a basis for designing frameworks concerned with the individual coastal areas. As a result, a *circular methodological track* arises including three steps.
a) *The building up of a priori general framework*. This classification is set up based on the literature on coastal management, existing coastal management programmes and other materials. Basically it is the result of a deductive approach.
b) *The use of the general frameworks as a guide to analyse and represent the use structure of an individual coastal area*. The general framework is utilised as a starting basis, to be related to the actual features of human presence and activity in the coastal area. As a result, a use framework referring to an individual coastal area is set up.
c) *The implementation of the general framework*. As the coastal area organisation evolves, its coastal use framework changes. Meanwhile, methodologies classifying the coastal uses progress. As a result, contributions arise from local experience

which may be used to implement the general coastal use framework. At this point the interaction between the *a priori* approach and the consideration of significant case studies comes to the fore.

As can be seen, this circular procedure makes it possible to optimise coastal management thanks to an endless co-operation between empirical and theoretical investigations. To be consistent with this objective, the general use framework should be modular and flexible so as to be easily implemented. This property could be operated if *nested classifications* were applied. In practical terms, the coastal use framework should not be articulated on a fixed number of ranks but on a variable number depending on the state of coastal exploitation and the degree to which it should be assessed. As an example, a tentative nested coastal use framework, derived from the development of categories included in Table 7.4, is presented in Table 7.5 for category 4 (*Living resource exploitation*) .

TABLE 7.4. A coastal use framework focusing on integrated management[1]

Categories of uses	Re: sustainable development components	Re: geographical components of the coastal area
1. Ecosystem conservation	IE	LD, BR, MR
2. Agriculture and livestock	EF	LD, BR
3. Natural vegetation use	IE, SE	LD, BR, MR
4. Living resource exploitation	EF	LD, BR, MR
5. Energy production	EF	LD, BR, MR
6. Mineral resource exploitation	EF	LD, BR, MR
7. Industry	EF	LD, BR, MR
8. Merchant seaports	EF	BR, MR
9. Sea navigation	EF	MR
10. Land transportation	EF, SE	LD, BR
11. Air navigation facilities	EF	LD, MR
12. Communication	EF	LD, BR, MR
13. Human settlements, urban and rural centres	EF, SE	LD, BR, MR
14. Recreation and tourism	EF, SE	LD, BR, MR
15. Cultural heritage	SO	LD, BR, MR
16. Research	IE, EF, SE	LD, BR, MR
17. Defence	IE, EF, SE	LD, BR, MR
18. Management	IE, EF, SE	LD, BR, MR

[1]   The table includes only the categories of uses, i.e. the first classification rank. Categories were defined bearing in mind the interaction between human presence and activities on land and at sea, and the subsequent impacts on the local ecosystem. Almost all the uses are developed through both man-made structures (buildings, railroads and others) and facilities (such as manufacturing plants). However, some of them are mostly developed through facilities so only this code is mentioned.

Codes:
Column 2: *IE*, Integrity of the ecosystem; *EF*, economic efficiency; *SE*, social equity.
Column 3: *LD*, Land site; *BR*, brackish site; *MR* Marine site.

TABLE 7.5.   Coastal use framework. Development of
category 4: living resource exploitation

Subcategory: 4.1—Fishing
4.1.1 Cash fishing
4.1.1.1 Demersal fishing
4.1.2 Trawling
4.1.2.1 Pelagic fishing
4.1.2.2 Environmentally-sound techniques
4.1.2.3 Explosive fishing
4.1.2.4 Cyanide fishing
4.1.3 Recreational fishing
4.1.3.1 Demersal fishing
4.1.3.2 Pelagic fishing
Subcategory: 4.2—Seining
4.2.1 Subsistence seining
4.2.2 Recreational seining
Subcategory: 4.3—Gathering
4.3.1 Seaweed gathering
4.3.2 Shellfish gathering
4.3.3 Purse gathering
Subcategory: 4.4—Clearing
4.4.1 Mangrove clearing
Subcategory: 4.5—Mining
4.5.1 Coral mining
Subcategory: 4.6—Aquaculture
4.6.1 Fish farming
4.6.2 Shellfish farming
4.6.3 Seaweed farming
4.6.4 Marine mammals farming
4.6.5 Seabird farming
Subcategory: 4.7—Hunting
4.7.1 Marine mammal hunting
4.7.2 Seabird hunting
4.7.3 Wild-fow hunting
Subcategory: 4.8—Transportation
4.8.1 Sea transportation
4.8.2 Land transportation
4.8.3 Fishing harbours
4.8.4 Fishing port terminals

7.5.   The Use-Use Matrix

After the coastal use framework has been constructed and its role in building up integrated coastal management has been clarified, the coastal use structure may be set up
using a square matrix. It includes, both in lines and columns and in the same order,
the elements classified by the coastal use framework. The nature of relationships between two uses is indicated at the crossing boxes between the relevant line and col-

umn. This breakdown, which can be called *Coastal use-use relationships* matrix, can be set up including only categories of uses, namely, the first rank of the coastal use framework, or can be extended to the second, third, and subsequent ranks. The main methodological issue is to determine how the relationships between uses are to be defined. Couper (1983: 208) adopted this classification:

a) harmful or conflicting interactions;
b) potentially hazardous interactions;
c) mutually beneficial interactions;
d) harmful to activity at matrix right (i.e., harmful to the use represented in the line);
e) beneficial to activity at matrix left (i.e., beneficial to the use represented in the column).

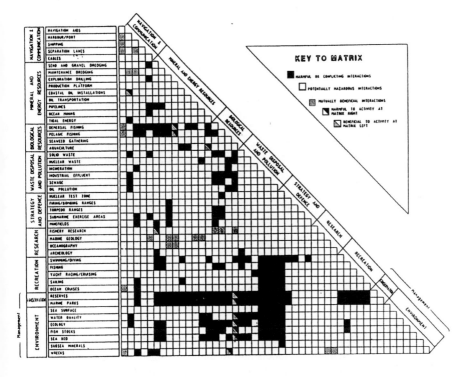

Figure 7.2. The Global Marine Interaction Model. Adapted from Couper ed. (1983: 208).

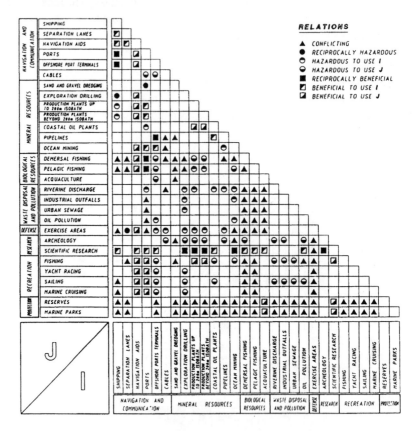

Figure 7.3. The Mediterranean coastal use-use relationships matrix.

The utility of this approach is due to the use of a square matrix, since it is an adequate tool to apply mathematical models and to map relationships, especially when the Geographical Information Systems (GIS) are applied. Nevertheless, to optimise the possibility of building up mathematical approaches and GIS-based assessment, it is useful for use-use relationships to be classified according to a binary logic-based approach. To this end it could be assumed that coastal uses are clustered in a square matrix where:

U_i (i ranged from 1 to n) are the uses of lines;
U_j (j ranged from 1 to n) are the uses of columns.
A binary logic-inspired approach could lead to this classification[35]:

1. *Absence of relationships*
2. *Presence of relationships*
    2.1.   *Neutral relationships*
    2.2.   *Conflicting relationships*
          2.2.1. For both U_i and U_j
          2.2.2. For U_i
          2.2.3. For U_j
    2.3.   *Beneficial relationships*
          2.3.1. For both U_i and U_j
          2.3.2. For U_i
          2.3.3. For U_j

Relationships are the result from inputs moving from one use to another, as well as crossing inputs. In this respect analysis could be based on three kinds of relationships according to input movements:

a) from U_i to U_j;
b) from U_j to U_i;
c) contextually from U_i to U_j and *vice versa*.

The case (i) implies that:

$$U_j = f(U_i)$$

namely, that the evolution of the use presented in the column is presumed to be influenced by the use presented in the line.

The case (ii) implies that:

$$U_i = f(U_j)$$

namely, that the evolution of the use presented in the line is presumed to be influenced by the use presented in the column.

The case (iii) occurs when:

$$U_j = f(U_i) \text{ and, contextually, } U_i = f(U_j)$$

namely, that the evolution of both the uses is reciprocally influenced.

Two comments seem appropriate.

First, *the patterns of relationship*. Cases (i) and (ii) appear as basically characterised by a *series relationship*. By its nature, case (iii), which is very frequent in coastal areas affected by high human pressure and a wide range of uses, is characterised by *feedback relationships*. As a result, the development of the retroactions could be a basic subject for assessing coastal use structures. Secondly, *the multiple use-caused interactions*. The more human pressure increases and the number of coastal uses grows the more numerous are the uses linked by relationships. As a result, the methodological approach should move from considering interactions between two uses to those involving more uses. On this subject it should be borne in mind that:

a) where a cluster of coastal uses is affected by *conflicting relationships* the whole coastal use structure is jeopardised by negative feedbacks;

b) where a cluster of coastal uses is affected by *beneficial relationships* the opposite occurs because there are positive feedbacks, and the coastal use structure benefits from the cohesion to pursue sustainable development objectives.

7.6. Conflicts between Coastal Uses: A Crucial Issue

With growing human pressure on the coastal areas and new types of coastal uses, there are increasing conflicts between uses. As a consequence, literature on coastal management has regarded conflicting relationships as the most intriguing and crucial issue of the use-use matrix. Therefore this subject will be dealt with by an *ad hoc* presentation in the following Chapter.

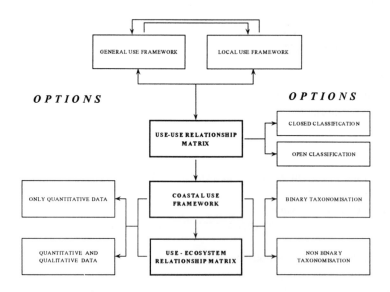

Figure 7.4. Procedure and options of analysis of the coastal use structure.

7.7. The Use-Ecosystem Relationships

In previous sections two steps of the presentation of the coastal use structure have been accomplished. They have been concerned with the coastal use framework and the use-use relationship matrix. Now, attention shifts to the third step, regarding the use-ecosystem relationships. As is well known, the relationships between coastal uses are fairly influential on those between uses and the ecosystem. Frequently, conflicting use-use relationships bring about stress to ecosystems. On the contrary, beneficial use-use relationships frequently mitigate the environmental impacts or are also healthy for the ecosystem. Moving from the conventional approach to a sustainable development-referred approach the need for assessing the use-ecosystem relationships increases and the relevant methodological concern becomes more complicated. As has been noted, the conventional approach—which had been developed during the 1970s and 1980s as a result of inputs from the *United Nations Conference on the Human Environment* (1972)—was based on the need to protect the environment *sensu lato*. In that context, attention primarily focused on pollution and was essentially concerned with the abiotic components of the ecosystems. The new approach, generated by inputs from UNCED (1992), fundamentally differs from the conventional one because it aims at optimising the management of the ecosystem. To be consistent with this point of view, a holistic framework of the interactions between coastal uses and the coastal ecosystem, or the set of contiguous coastal ecosystems, needs to be provided. This need may be met by setting up the *Coastal use-ecosystem relationships matrix* consisting of a rectangular matrix where:

- in *lines*, the use U_i (where i ranges from 1 to n) are presented in the same order which was adopted in the *coastal use-use relationship* matrix;
- in *columns*, the components E_j (where j ranges from 1 to m) of the coastal ecosystem are presented;
- in *boxes,* at the crossing of lines and columns, the impacts from the coastal uses on the coastal ecosystem are indicated.

The most critical issue relates to the criteria with which the ecosystem is shown in columns. As a preliminary approach, it seems useful to evaluate the utility of:

a) keeping the biotic elements distinct from the abiotic elements of the ecosystem;

b) taking into consideration those properties of the ecosystem which are concerned with its integrity;

c) framing these cognitive elements into the biogeochemical cycles, as they are dealt with in the International Geosphere-Biosphere Programme (general framework) and GESAMP (ocean management);

d) bearing in mind that the coastal area, extending from land to the sea, embraces various kinds of ecosystems.

As an example, the basic design of columns could be as follows:

a) *biogeochemical cycles and processes* (carbon dioxide cycle, hydrological cycle, erosion cycle, sedimentation processes, subsidence, metal cycles, nutrient cycles, and others);

b) *abiotic components and properties of the coastal ecosystem* (geomorphologic features, sediments, physical and chemical water properties, water dynamics, and others);

c) *biotic components and properties of the coastal ecosystem* (land fauna and flora; land, brackish water and marine trophic webs; and others).

d) *properties of the ecosystem* (productivity, resilience, biodiversity and others).

Only some relationships between the coastal uses and the ecosystem may be measured. As a result, both quantitative and qualitative data are included in the crossing boxes of the matrix. Where possible, the impacts from sets of use should be highlighted and measured. This need makes the methodological concern more complicated.

7.8.   The Coastal Use Structure Facing Options

Investigations on the coastal use structure have three basic roles:
a) a *cognitive role*, leading to the assessment of coastal uses, the use-use relationships, and the use-ecosystem relationships;
b) a *prospective role*, based on the building up of scenarios aimed at designing possible future coastal use structures;
c) a *management-oriented role*, aimed at designing the most suitable coastal use structure for the individual ICM programme.

Following this approach, the matrices illustrating the existing coastal use structure (cognitive role) may be regarded as bases for:

• setting up *hypothetical matrices*, reflecting possible future use structures, rather different from the existing one, each of them characterised by its own impacts on the coastal ecosystem and able to support a single development path (prospective role);

• selecting the *optimum matrix* which maximises possibilities of achieving sustainable development (management-oriented role).

The hypothetical matrices lead to designing scenarios while the optimum matrix is normative and planning-oriented, and leads to designing the ICM programme. It goes without saying that the latter is drawn by the former. To follow these routes, it should be enquired *which kinds of coastal uses* are:
a) becoming obsolete and are going to be left;
b) expected to be left because they are regarded by coastal community and decision-making centres as not fitting for coastal development;
c) expected to be left because they bring about impacts on the coastal ecosystem or cultural heritage which cannot be tolerated.

This operation is to be associated with another conceived to identify *what new kinds of coastal uses are expected to be located in the coastal area because they are more*:
a) technologically advanced *vis-à-vis* the existing coastal uses;
b) suitable to guarantee the protection of the local ecosystem;
c) qualified to protect the local cultural heritage;
d) fitted to optimise scenic values.

Where attention is centred on the design of use-ecosystem relationships consistent with the pursuit of sustainable development (optimum coastal management pattern) it is necessary to take into account a range of basic options which the coastal community and decision-making systems usually deal with. While exploring literature on ICM, the following options are worthy of consideration.

Ecosystem versus *mere pollution management:*
decision making-centres (a) aim at innovating environmental policy assuming the pursuit of ecosystem's integrity as the fundamental goal, or (b) they persist in simply keeping pollution under control.

Consumptive versus *non consumptive economy:*
economic system (a) is driven in such a way as to minimise the consumption of natural resources, or (b) it persists in developing conventional behaviour maximising their use.

Renewable resource versus *non renewable resource exploitation:*
decision-making centres (i) develop efforts to substitute the use of non renewable resources with that of renewable ones, or (ii) they persist in maximising the use of non renewable energy and mineral resources.

Social versus *individual:*
coastal uses are managed with the aim of (i) optimising social needs, or (i) assuming, as a main principle, the satisfaction of individual needs.

Social versus *commercial:*
local economy is driven in such a way as (a) to develop non profit-aimed social facilities and utility industries, or (b) to stimulate profit-making uses.

Coastal dependent versus *non coastal dependent use development:*
the local economy (i) aims at becoming increasingly coastal dependent, namely, consisting of an increasing range of uses requiring "an immediate coastal site to be able to function at all", or (ii) it is based on a wide range of non-coastal dependent uses (Sorensen and McCreary 1990: 13).

Offshore-oriented location versus *onshore-oriented location:*
the coastal uses (i) progressively gravitate to the marine environment through the establishment of offshore facilities and marine activities, or (ii) they are mainly based on land facilities and activities.

Present needs versus *future needs:*
coastal community (i) is aware of the need to protect the rights of future generations, or (ii) it persists on a conventional, present generation-referred, management of natural and cultural resources.

As can be seen, some options are directly concerned with the impacts on the coastal ecosystem from human presence and activities, while others are only indirectly concerned. Nevertheless, it is essential to design a comprehensive view of these impacts with the aim of understanding what evolution is addressed to the ecosystem because of the coastal organisations.

Environmental Impacts from Living Resource Exploitation. Notes from OECD
As far as environmental impacts from coastal uses are concerned, attention has usually been paid to those brought about by manufacturing plants, seaports, settlements and other uses generally concerned with secondary and tertiary sectors of the local economic system. However, living resource exploitation can also cause remarkable impacts on the coastal ecosystem. On this subject OECD noted:
    "(An important issue) is the effect of pollution caused by finfish farms. This is still a new area of study and thus the impact of this pollution on the marine ecosystem is difficult to establish. Many factors affect the dispersal of fish farm wastes, such as water depth, current speed and direction, sea bottom topography, type of cage mooring used, seasonal variations in volume or river flow. It has been argued that finfish, with their need for unpolluted water, would be the first to suffer from waste pollu-

tion. The presence of finfish is often considered a "clean water indicator". It has been pointed out, however, that salmon can thrive in much higher concentrations of nutrients and certain other chemicals than would be considered desirable for the general ecological health of sea lochs".
(From OECD [1993: 99]).

7.9.   The Coastal Use Structure Facing ICM

The developed reasoning and the subsequent methodological ways presented in this Chapter highlight both the complexity of the coastal ecosystem—where it is conceived in accordance with ICM and sustainable development concepts—and the need to strengthen efforts to produce suitable approaches. It is also clear that the methodological way requires additional steps, at least aiming at:
a) providing criteria and parameters capable of measuring the consistency of individual coastal uses, or groups of uses, with the sustainable management of the coastal system;
b) designing criteria through which the properties of the biotic communities—especially resilience, productivity and biodiversity—may be assessed and measured in accordance with the need to preserve the integrity of the ecosystem from the inputs deriving from the use structure;
c) mapping the coastal use structure through criteria tailored to the need to evaluate its consistency with the sustainable development principle;
d) in particular, providing maps focusing on the global impacts that the coastal use structure as a whole brings about on the coastal ecosystem;
e) integrating remote sensing and other source-based information, and properly presenting them through multi-media tools, such as GIS.
There is no doubt that holistic investigations into the coastal use structure are no easy task because they need to build complicated matrices and to reason by processes and interaction. Matrices provide a basis for discussing the existing (cognitive level), possible (prospective level), and optimum (normative level) coastal organisation. Furthermore, they introduce coastal management to the use of GIS.

Experiments carried out in pilot educational courses[36] have noted that when attention focuses on the individual coastal area, the basic difficulty is met in designing the coastal use framework. Many questions come to the fore regarding whether and how facilities, activities, and others, should be classified and on how many levels the framework is to be based. Opposite to what might be expected, minor difficulties are encountered when the use-use relationships matrix is set up. However, the situation is even more difficult when the use-ecosystem relationships matrix is constructed, as it is concerned with how impacts should be classified and mapped. This is primarily due to the lack of contributions from ecology. In spite of these methodological difficulties, classifications, clustering and mapping lead to perceiving the coastal use structure in ways consistent with the need to design useful holistic views for ICM programmes.

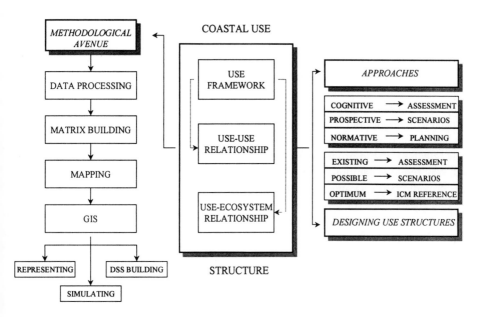

Figure 7.5. The basic aspects and outcomes of the methodological approach to the coastal use structure. GIS: Geographical Information Systems; DSS: Decision Supporting System.

Mapping data with matrices (cognitive, prospective and normative) is a preliminary step for setting up coastal GIS to focus on the coastal system organisation. This issue appears to be merely technical but it is actually concerned with how the coastal structure should be conceptualised and what methodological procedures should be applied. This concern is quite new in the scientific arena, and it has a key role with regard to how ICM is carried out. To optimise approaches two parallel exercises are possible. On the one hand, approaches to new coastal area programmes may follow such innovative procedures as that presented in this Chapter to describe the coastal structure and system holistically. Without holism-based procedures, the prospect of building up ICM is frustrated and seriously jeopardised. On the other hand, the procedure may be applied to existing coastal management programmes to explore whether and how they would have been formulated differently if the coastal use structure had been investigated holistically. The Mediterranean offers a suitable ground for carrying out this exercise. The existing Coastal Area Management Programmes (CAMPs) of MAP (Table 1.8) make it possible to perform the latter exercise, while the third generation of CAMPs, which is going to be dealt with, can conduct the former exercise. The final aim is to combine conceptual and empirical research to improve methods and advance coastal management programmes.

CHAPTER 8

CONFLICTS BETWEEN COASTAL USES

This Chapter will focus on the nature of conflicts between coastal uses and their role in the context of management issues. Approaches to assessing and managing conflicts concerned with the need to operate integrated coastal management programmes will be designed and proposed. The following questions will be discussed:

Questions	Figures	Tables
How are conflicts between uses relevant to ICM?	8.1	8.1
How may multi-national conflicts be approached?	8.2	
Which parameters can be used to assess conflicts between uses?		8.3, 8.4
Which parameters can be used to assess conflicts between users?		8.2 to 8.6
How may the causes of conflicts be approached?	8.3	
How tractable are the conflicts?	8.4	8.5
How may a holistic view of conflicts between uses be set up?	8.5	8.6
What code of conduct regarding conflicts between uses may be adopted by managers and planners?	8.5	

8.1. Focus on Conflicts

The previous Chapter showed how the coastal use structure may be assessed, and proposed a matrix-based model investigating the relationships between uses. The model was designed to distinguish the beneficial and conflicting relationships and to distinctly focus on each of these categories. This distinction dates back to the first structuralist approach (Couper ed. 1983) to coastal use structure, and has since become increasingly important for assessment and management purposes for two reasons.

First, where there is a conflict between two or more different uses—e.g. between aquaculture and thermoelectric plants, or between industry and tourist harbours—impacts are often addressed to the local ecosystem too. They may be really stressful, for example when opting between installing a petrochemical plant and a recreational harbour the local decision-making system chooses the plant, thereby altering the abiotic niche of the ecosystem and influencing the organisation of its biotic community. In other cases, impacts alter the biotic community directly. For instance, when a conflict arises between a thermoelectric plant and a fishing harbour and the plant is preferred, warm water is disseminated into the local marine water column. This brings about a rise in thermal energy, thereby accelerating the chemical and biological processes of the biotic community.

163

Secondly, conflicts between uses frequently have impacts on the economic and social organisation of the coastal area since, according to the preferred use, the local
system undergoes specific inputs and a specific pattern of natural resource use solidifies. In this respect, looking at the individual coastal area, it is necessary to evaluate
how and to what extent the prospect of pursuing sustainable development could be influenced by conflicts between uses and their possible outcomes. As regards the conflict between the petrochemical plant and the recreational harbour, the former would
definitely drive the local economic system towards the primary production industry,
while the recreational harbour would lead it towards the tertiary sector. Furthermore,
different patterns of resource exploitation arise. Petrochemical plants require vast
spaces, fresh water, and land transport infrastructures, whereas recreational harbours
need limited land areas, access to, and availability of, marine water surface facing the
coast, and aesthetically pleasant land facilities.

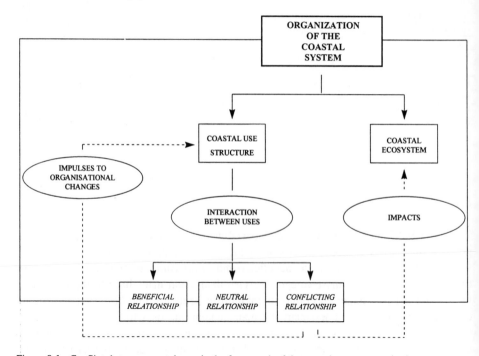

Figure 8.1. Conflicts between coastal uses in the framework of the coastal system organisation.

Where conflicts between uses are analysed with the aim of operating ICM, it is
important:
a) to deal with the conflicting uses adopting a well-defined concept of conflict;
b) to use a methodological approach capable of understanding the relevance of conflicts to coastal sustainable development;
c) to delineate a view of methods useful to prevent, mitigate and solve conflicts.

8.2.   The Multi-National Scale: International Conflicts

The approach employed to assess conflicts between uses acquires very different features according to the scale concerned, and so two key scales need to be distinguished. On the one hand, the multi-national scale arises, which is concerned with international legal frameworks—viz. international and regional conventions, multilateral and bilateral treaties, and so on—and jurisdictional organisations, such as the International Court of Justice. On the other hand, the national scale arises, which is concerned with the national legal framework, consisting of the juridical discipline and provisions from the federal and state governments, as well as subnational branches of public authorities.

Three basic kinds of conflict can be found on the multi-national scale:

a) *jurisdictional conflicts*, often concerned with the national maritime jurisdictional zones and their boundaries;

b) *environmental conflicts*, frequently concerned with pollution or, *sensu lato*, ecosystem degradation;

c) *economic conflicts*, frequently referring to the exploitation of coastal resources.

The *jurisdictional conflicts* occur where opposite or adjacent states claim rights on the same maritime zone (Birnie 1987). Conflicts on territorial seas and continental shelves have arisen and diffused in recent decades, especially in enclosed and semi-enclosed seas. As regards the Mediterranean, the conflicts between Greece and Turkey—which were presented in Chapter 4.3—although legally grounded on the baselines, are brought about by conflicting economic interests on the territorial sea and the continental shelf (Kliot 1987 and 1989). Where conflicts are concerned with the territorial sea, they usually involve navigation, overflight and coastal fisheries. Where they are concerned with the continental shelf, they usually apply to oil and gas exploitation and exploration. The conflict in all these cases regards the same use—navigation, fisheries, oil and gas exploitation—that the two states want to practise in the same marine space where both states claim their jurisdiction.

Different features can be found in *environmental conflicts* because one state usually harms the spatial environment—terrestrial, atmospheric, marine—where another state's jurisdiction extends. Pollutants may cross boundaries through atmospheric circulation, rivers, aquifers and marine currents. Potential or actual Mediterranean conflicts arise between adjacent states because pollutants are discharged into the coastal sea by one state, are carried by marine currents to the another state's coastal waters, and consequently damage their fisheries and jeopardised their recreational uses. These conflicts have frequently occurred between France and Italy: pollutants discharged into the Ligurian Sea have been transferred to the marine waters facing the Côte d'Azur, which is characterised by a wide web of coastal uses and high coastal human pressure. In 1996, the member states of the Barcelona Convention adopted a protocol on transboundary movements to prevent this kind of conflicts.

Economic conflicts between states occur where one state practises uses that are claimed as infringing international law by another state. One of the most common cases concerns fisheries where one state allows fishing in waters that another state claims as part of its jurisdictional zone, and is therefore charged with overlapping national waters. For the Mediterranean, such conflicts have occurred between Tunisia and Italy on numerous occasions when Italian fishing vessels have been charged with entering Tunisian territorial waters to fish. In actual fact, conflicts over fishing are the most frequent in the world and, without any efficient legal tools and political agreements between states, they are expected to spread further as demand for living marine resources grows. Most literature on conflicts between states is concerned with this issue.

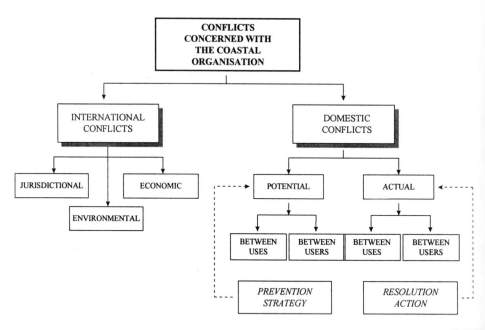

Figure 8.2. International and domestic conflicts between coastal uses.

8.3.   The Local Conflicts: Focus on Uses

The multi-national scale of conflicts is closely relevant to coastal management. The manager and planner are interested in these conflicts where the coastal area considered by the ICM programme is opposite or adjacent to another state's coasts. Shifting from the multi-national to national scale, a wide range of domestic conflicts arises (Cicin-Sain 1992: 281). There are not only many more conflicts than international ones, but they are also closely turned to the focal aspects of coastal organisation. As a consequence, they are particularly relevant to ICM. These conflicts may be considered in the light of both coastal uses and coastal users. The approach to coastal uses leads to evaluating how the coastal resources—e.g., the biomass of salt and brackish waters—are exploited, which technical tools are employed, what organisation has been set up, what environmental and social impacts have been brought about, and so on. The nature of use should be highlighted rather than who practises it. This approach is rather abstract and occurs more in European literature than in American literature, which is more interested in the concrete behaviour of users—namely, firms, local authorities, governmental authorities etc.—having a great influence on the specific conflicts. The best approach might be to focus first on the uses and their means of ex-ploiting coastal resources, and then on users and the social and economic factors causing the conflicts, such as firms' strategies and decision-makers' attitudes.

An *a priori* web of conflicting uses was presented in Chapter 7, where the *coastal use-use relationships* model was considered. As already stressed, the model simply provides ground for discussing the general aspects of conflicts. To proceed along this methodological route it should be noted that the conflict relates to the use or non use of a particular part of the coastal area (Cicin-Sain 1992: 281; Cicin-Sain and Knecht 1998: 25-35). When the option between different use patterns of a given space arises—such as between a bathing or an aquaculture plant, a shipyard or an iron and steel plant— the social community and the local decision-making system are often di-vided into two groups. Each group is willing to adopt one of the options in order to direct the local economy towards a given goal, to encourage a given environmental approach, and so on. Frequently the option is basically concerned with whether the local ecosystem, or parts of it, is to be conserved or changed. In this case, a conser-vation *versus* exploitation conflict arises somewhat involving the social perception of the coastal management's final goal, and the rights of future generations. As an even minor conflict between uses effects the cultural roots of local communities, the use-use relationships model may be referred to again to have a better understanding of the complicated framework of consequences caused by conflicts. In this respect, the fol-lowing aspects may be worthy of attention.

8.3.1. THE CONFLICT VECTOR

The conflict between uses can be represented by *ad hoc* matrices and maps, and imagined as one impulse moving from one individual coastal use to another, so-called a mono-direction vector. More complicated patterns may arise consisting of two si-multaneous or sequential impulses moving from one use to another, and *vice versa*, so-called a bi-direction vector. An example of the mono-direction vector can be found in inputs moving from a recently-built loading or unloading seaport terminal for liquid bulk to a bathing place nearby. The former causes environmental impacts affecting the visual landscape features. The latter cannot tolerate this, and hence there is incom-patibility between the two uses. The bi-direction vector may be exemplified in the conflict between a bulk seaport terminal and a containerised seaport, where both are located in a coastal area where maritime transportation develops. The former tends to enlarge using new coastal spaces while the latter needs wider spaces for handling and logistic operations. Inputs move from the former to the latter, and *vice versa*, leading the local decision-making centres to choose between two different paths for the local seaport economy: (i) optimise the interaction between the seaport and manufacturing

activities; or (ii) optimise the interaction between the seaport and the national and international networks of logistic platforms.

8.3.2. THE NUMBER OF CONFLICTING USES

Literature has concentrated on conflicts between two uses. Nevertheless, as human
pressure on coastal areas grows and the web of uses widens, conflicts between more
than a couple of uses spread. In relation to the seaport, conflicts between urban waterfront uses, seaport container terminals, and industrial facilities served by bulk seaport
terminals are spreading out in many countries, as a consequence of economic development of coastal regions. The EU side of the Mediterranean, for example the Gulf of
Genoa, is influenced by such a process.

8.3.3. DYNAMIC VIEW

It is normal for the individual conflict to be considered in relation to those features
characterising it at a given moment. However, the conflict is a *process* acquiring different aspects and bringing about different impacts during its evolution. As a result,
coastal management could benefit from a dynamic view when choosing the best options for the conflict.

8.3.4. POTENTIAL AND ACTUAL CONFLICTS

Conflicts may arise between:
a) two potential uses of the same resource, e.g. between the prospects of locating an
    aquaculture plant or a recreational harbour in a given coastal place;
b) an existing use (e.g., an existing iron and steel plant) and a potential use (e.g., a
    planned technological park) involving the same place;
c) two existing uses (e.g., the mentioned conflict between the bulk and containerised
    terminals on the same seaport area).
    Potential conflicts greatly influence coastal management, particularly in the design
of management programmes. If they are identified and assessed in advance, measures
may be adopted to prevent them and management may be highly effective. For example, many potential conflicts are concerned with the implementation and diffusion of
fish farming and other aquaculture facilities.

Conflicts Concerned with Living Resource Use, According to OECD
Many conflicts have arisen between the different uses concerned with living resource exploitation. This
was emphasised by OECD which considered many cases, and the following are worth mentioning.

Focusing on Countries
Canada. "Particular conflicts are identified as those between aquaculture, traditional fisheries and recreation. In the Fraser River estuary, for example, apart from the conflicts generated by pollution and
habitat damage, several others had become increasingly evident in the early 1970s including *inter alia*:
*   conflicts between commercial, sport, and native fishing interest and within the commercial fishery
    between gear types;

- recreationalists conflicted with each other as new demands emerged (e.g. off road vehicle use on the dunes and sand bars);
- more and more agricultural land was being sold to developers;
- congestion as increased numbers of recreational boaters sought moorings and competed for space with other port and industrial activities (e.g. booming, port expansion)".

Norway. "Aquaculture is a major growth industry (...) characterised by a pattern of many relatively small firms. (....) In many regions fish farming has become the dominant user of available water space within the last 10 years. This has caused several conflicts with other user groups such as traditional fishermen, recreation and conservation interests".

Spain. "Tourism is a major activity in Spain's southern and eastern coasts. This, in combination with the urbanization of the coast and development of infrastructure, has contributed to some conflicts with nature conservation and maintenance of coastal processes".

Focusing on Aquaculture
As long as there is saturation of catchable marine living resources, aquaculture is expected to become increasingly important in both developed and developing countries. However, this use is liable to conflict greatly with other coastal uses. OECD (1993: 99) notes:
"Aquaculture developments are becoming increasingly common in coastal areas where relevant criteria are satisfied, such as water quality, sheltered location, ease of access, close proximity to loading facilities, transport networks, processing plants and/or market. Such developments provide economic and social benefits in terms of employment, direct and indirect (multiplier) revenue generation, and opportunity for rural development and diversification. The activities of externalities of coastal land and water uses can affect the 'environmental health' of an aquaculture development. Inputs into harbours, bays and estuaries of pollution from industrial and commercial landuses, runoff from agricultural land and sewage effluent discharge are obvious examples".
From OECD (1993a: 92–9).

8.3.5. PHILOSOPHICAL AND IMAGINED CONFLICTS

Both potential and actual conflicts are based on facts since they refer to a competition between two or more existing uses, namely, something which has solidified, or it is going to take place, in the coastal area. On the other hand, there are conflicts not based on facts, although they generally influence coastal management. According to Cicin-Sain (1992: 290–1), these conflicts can be categorised as philosophical and imagined. Philosophical conflicts "involve value differences which usually occur between indirect and direct users. These philosophical conflicts are most pronounced when parties deny that other parties have legitimate concerns to be considered; when for example, environmentalists hold that oil operators should simply 'not be there' and oil industry opposes the environmentalist's right to a say in marine operations". Imagined conflicts are concerned with perception and "may be rooted more in differing perceptions of facts and probabilities of impacts, perhaps because of poor communications among interests, and may be resolved more easily by improved data collection and information sharing".

8.3.6. EXISTING AND NEW TYPES OF USE

Conflicts have a different relevance to management according to whether they have arisen between existing uses, or between an existing and a new, possible or expected type of use characterised by innovative organisation and technical features. The former case is a characteristic of coastal areas not undergoing structural changes and might denote conservative regional policy. Usually the latter case occurs in coastal areas subject to structural changes and denotes regional policy aimed at innovating economic organisation.

8.3.7. RELEVANCE TO SUSTAINABLE DEVELOPMENT

Analysing conflicts between uses cannot disregard their relevance to the final objective of coastal management programmes. Firstly, the management's final goal, i.e. sustainable development, has to be focused on. Then the role of conflicts facing the coastal area organisation should be considered by evaluating whether and how the individual conflict jeopardises the prospect of guaranteeing the ecosystem's integrity, pursuing economic efficiency and protecting social equity, including the rights of future generations. The more the conflict jeopardises this prospect, the more management should be keen on adopting measures to mitigate it.

8.3.8. RELEVANCE TO THE SPECIFIC GOAL

After evaluating the relevance of conflicts to sustainable development, attention shifts to the individual goals of coastal area management, such as development of recreational uses and optimisation of living resource exploitation. When conflicts jeopardise the pursuit of individual goals they need to be dealt with by special prevention and resolution mechanisms. The associated analysis of conflicts' relevance to both final goal (sustainable development) and specific goals leads us to the *teleological appreciation* of conflicts.

8.4.   The Local Conflicts: Focus on Users

As has been mentioned, conflicts may be considered referring to the use or to the user. The latter approach leads to take into account behavioural patterns and strategies adopted by decision-making systems to locate, implement and manage uses. It is complementary to the former approach, since after considering the nature of conflicts attention logically shifts to bodies, such as local authorities and entrepreneurs, involved in conflicts. To introduce this subject it is worth bearing in mind that, in the literature on coastal management, the word "user" has acquired two different meanings, because it is intended as:

a) the body—viz. a firm, an organisation, a company, a holding—exploiting one or more coastal resources (e.g., iron and steel industry using littoral space for seaport

terminals and processing plants; fisheries exploiting living resources, a firm managing a marine museum and so exploiting cultural heritage);
b) the body—namely, a state department, a state agency, a municipal corporation—which has jurisdiction over an individual use (e.g., the Ministry of Agriculture having jurisdiction over marine living resource exploitation), or a set of uses (e.g., the Ministry of Transportation having jurisdiction over both maritime and land transportation).

In the first case, conflicts are concerned with coastal uses and consider "which of the conflicting uses should be preferred" or "how the conflicting uses may coexist". In the second case, conflicts become "disputes" concerned with the public approach to coastal resource exploitation and consider "which authority has jurisdiction over the individual use and user". Both questions are closely linked to the consistency of coastal management with the principles of sustainability and ICM, and so the way they are approached influences the effectiveness of coastal management. As regards public administration, conflicts involve bodies with jurisdiction over uses (e.g. a Ministry department involved in oil and gas exploration and exploitation) and bodies practising uses (e.g., a state agency installing and managing oil and gas plants).

There are many ways to assess the nature and role of users. As a preliminary approach, the user and its role may be identified by three basic areas:
a) *Legal status*, where users can be grouped into three categories: *public users*, whose actions are regulated by public law and kept under control by public law (e.g., a state agency on oil and gas exploration and exploitation); *private users*, whose actions are regulated by private law (e.g., a private company operating in oil and gas exploitation); and *combined users*, who act in the public's interest but are regulated by private law.
b) *Sectoral Operational area*, where users are grouped according to the classification of uses (e.g., seaport authority, manufacturing company, fishery, etc.).
c) *Geographical operational area*, where users are grouped according to the geographical area they operate on. The public administration, for example, may be grouped federal, state, regional (subregional) and local bodies.

On this basis, the nature of conflicts between users may be investigated and its relevance to ICM be evaluated. Cicin-Sain (1992: 281–3) points out that there are two main categories of conflicts, respectively between:
a) "users over the use or non-use of particular ocean areas; and
b) government agencies which implement marine laws and policies".

Users are direct and indirect. "*Direct* users include those who actually physically participate in marine activities (such as recreational and commercial fishermen, oil operators),as well as parties with allied roles (such as bait suppliers, fish processors). Examples of *indirect* or potential users include environmental groups which promote and protect the non-utilitarian values of the ocean, members of the public who do not directly utilise the ocean or who live in areas remote from the coastline, and future generations" (Cicin-Sain 1992: 282).

Conflicts between agencies could involve agencies on the same operational level (*inter-agency conflict*) or agencies on different operational levels (*inter-governmental conflict*). As a conclusion, the approach from Cicin-Sain leads to the classification presented in Table 8.1.

TABLE 8.1.   Types of conflicts between users[37]

1. Direct user versus direct user	
1.1	on the establishment or implementation of a particular use;
1.2	on the non-establishment or dismissal of a particular use.
2. Indirect user versus direct user	
2.1	non use *versus* the use;
2.2	public property-sustaining social groups *versus* direct users;
2.2.1	local social groups *versus* direct users;
2.2.2	remotely-located social groups *versus* direct users;
2.3	environmental-sound social groups *versus* direct users;
2.4	future generations *versus* direct users;
2.5	others.
3. Governmental agency versus governmental agency	
3.1	agencies of the same governmental level;
3.1.1	inter-agency conflicts;
3.2	agencies of different governmental levels;
3.2.1	inter-governmental conflicts.
Adapted from Cicin-Sain (1992).	

The classification given by Cicin-Sain is basically inspired by the United States system and subsequent experience in coastal zone management. Vallega (1993: Chapter 8) provided a framework sensitive to the EU context with special reference to the Italian legal framework and governmental organisations, which has largely been influenced by the "Code Napoléon" legal system. The relevant framework is based on these concepts:

1. *user* is intended as a private user exploiting (*actual user*), or going to exploit (*potential user*) an individual coastal resource or a set of resources;
2. *public administration* is intended as embracing:
    2.1. the central government consisting of departments (ministries) and their local branches;
    2.2. governmental agencies, e.g., a national organisation managing a public sector, such as railways or public works;
    2.3. local government consisting of autonomous:
       2.3.1.   regional authorities, such as *Länder* (Germany), *Départments* (France), *Regioni* (Italy); and
       2.3.2.   local authorities, such as municipalities.

These concepts lead to designing the classification presented in Table 8.2.

TABLE 8.2.   Types of conflict between users: a view of the European context

Types of conflict	Examples
1. User *versus* user	
1. One user versus another user.	Offshore oil and gas versus navigation
2. Intra-governmental conflicts	
2.1 Department versus department	Ministry of Agriculture versus Ministry of the Environment on the location of a coastal railway
2.2 Local departmental branch versus local departmental branch	Local branch of the Ministry of the Environment versus the local branch of the Ministry of Public Works on the location and physical design of a seaport breakwater.
3. Government *versus* agencies	
3.1 A governmental department versus a governmental or intergovernmental agency	Ministry of Industry versus Energy Agency on the national programme of coastal thermoelectric plants
4. Agency *versus* agency	
4.1 A governmental agency versus another governmental agency	Telecommunication agency (submarine transmission cables) versus Energy Agency (offshore oil and gas installations)
4.2 A governmental agency versus an inter-governmental agency	Electric Energy Agency (thermoelectric plants discharging warm water into marine coastal waters) versus FAO and UNEP.
5. User *versus* public sector	
5.1. User versus department	Fisheries versus Ministry of Agriculture on the aquaculture development programme
5.2. User versus a local branch of department	Bathing plant versus the local branch of the Ministry of Public Works on the location of a recreational harbour
Adapted from Vallega (1993: Chapter 8).	

8.5.   Causes of Conflict

When integrated coastal management is operated on the local scale, it is necessary to shift from a global view of conflicts between uses, and the subsequent impacts on the ecosystems, to the individual conflicts focusing on its nature, possible evolution and resolution, or—where the conflict is latent—on its possible prevention. To deal with this need, the reasons by which it was triggered and solidified should be analysed. In this respect, different criteria may be adopted according to the conceptual and methodological approach adopted to deal with coastal management, and the local socio-economic context, but whatever criterion is adopted, a classification of conflict motivation is to be defined as a basis for framing the individual cases. By way of example, a classification of conflict roots is shown in Table 8.3. Only four roots are considered in order to simplify the approach to an issue, which is methodologically complicated, and politically demanding.

TABLE 8.3. Main types of incompatibility between coastal uses

Incompatibility	Reason	Example
Locational	Two or more uses need to be located in the same place but there is not enough room for all of them.	Conflict between naval exercise areas *versus* mercantile navigation.
Organisational	One use is organised in terms damaging other use(s).	Navigation (supply vessels and others) serving offshore oil and gas fields *versus* yachting and cruising.
Environmental	One use has impacts on the local ecosystem and therefore damages other use(s) based on the ecosystem's conservation.	Marine sanctuary *versus* thermoelectric plant discharging warm water.
Visual	One use alters scenic values in terms that other use(s) cannot tolerate.	Heavy manufacturing plants, such as iron and steel plants, *versus* recreational facilities.

8.6.   Relevance of the Jurisdictional Zones

There is no doubt that the tractability of conflicts between coastal uses, and coastal users, depends on many factors, and acquires different features according to which potential and actual conflicts are considered. In this perspective, the social conditions and the legal framework play key roles. As regards the former range of factors, that is social conditions, each coastal areas is marked by its own features depending on the economic level, quality of life, ethnical and cultural stress, and others. Moreover, in any place the tractability of conflicts is more or less influenced by the social perception of the need for coastal management programmes, and the social participation level in decision-making processes. When attention shifts to the legal framework it should be taken into account that each coastal area is influenced, landwards, by the property regime and administrative organisation, and seawards by the local jurisdictional regime consisting of a set of jurisdictional zones, each of them characterised by its own legal framework. As a result, the understanding of the conflict's nature and tractability requires first the location of the individual conflicting uses to be identified. In this perspective, the *legal status of use location* becomes essential in understanding the role of conflicts with regard to coastal management. Regardless of the framework of jurisdictional zones presented in Chapter 4, three main spatial extents may be taken into account::
a) the *land coastal areas*, which are subject to national law and are characterised by the property regime;
b) the *national internal waters*, which have the same legal status as land and are characterised by the property regime;

c) the *jurisdictional maritime zones*, which are regulated by international conventions ratified by the state and are not characterised by the property regime but are only subject to the state's jurisdiction.

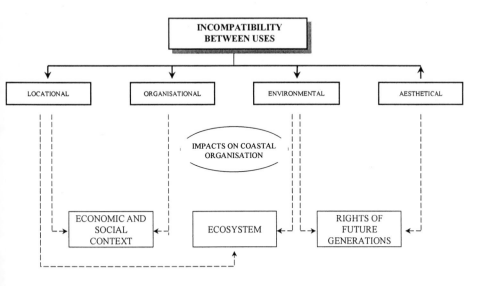

Figure 8.3. Main types of causes of conflict between uses and main possible impacts.

Bearing this classification in mind, the breakdown presented in Table 8.4 has close relevance to the prospect of carrying out ICM programmes.

TABLE 8.4. The legal status of coastal areas

Legal status	Near atmosphere	Sea surface	Water column	Seabed	Subsoil
1. Land	NL	NL	NL	NL	NL
2. Internal waters	NL	NL	NL	NL	NL
3. Jurisdictional maritime zones					
3.1 Territorial sea	NJ	NJ	NJ	NJ	NJ
3.2 Archipelagic waters	NJ	NJ	NJ	NJ	NJ
3.3 Contiguous zone [(1)]	IR	NJ	IR	IR	IR
3.4 Continental shelf	IR	IR	IR	NJ	NJ
3.5 Exclusive economic zone	NJ	NJ	NJ	NJ	NJ
3.6 Exclusive fishery zone	IR	NJ	NJ	IR	IR

Codes:

NL national law: the physical component of the zone is subject to the national law;

NJ national jurisdiction: the physical component of the zone is subject to a ratified convention;

IR international regime: the physical component of the zone is under international regime.

[(1)] The state's authority over the contiguous zone is only concerned with the infringement of its customs, fiscal, immigration and sanitary laws (LOS Convention, Art. 33).

8.7.   Tractability and Management of Conflicts

Apart from the distinction between direct and indirect users, Cicin-Sain (1992: 290–3) believes that conflicts can be categorised according to their *roots* (differences in values, interests, and facts). Combining the classification based on roots and being aware that conflicts may be categorised as philosophical and imagined, potential and actual, their tractability may be considered. It is to be intended in an orthodox sense, that is, capable of being led, taught or controlled. With regard to conflicts, it may be defined more specifically as the attitude, or capability of being prevented, mitigated, controlled, solved, removed. Tractability is decreasing, shifting from imagined to actual, potential and philosophical conflicts. The relevant breakdown is given in Table 8.5.

TABLE 8.5.   Tractability of conflicts

	Philosophical Conflict	Potential conflict	Actual conflict	Imagined conflict
Roots of conflict	differences in values	facts, interests, possible values	differences between facts, interests	differences between perception of the fact
Parties to conflict	indirect users/direct users	Direct users/direct users	direct users/direct users	direct users/direct users
Degrees of Tractability	Lowest tractability	Low Tractability	Medium tractability	High tractability

Adapted from Cicin-Sain (1992: 292; see also Cicin-Sain and Knecht, 1998, 234).

To manage conflicts between uses, or between users, reactive and proactive conflict management can be undertaken. The former approach aims at addressing and resolving conflicts that have already materialised; the latter aims at preventing conflicts that are expected or supposed to materialise.

8.7.1. REACTIVE CONFLICT MANAGEMENT

Two main kinds of reactive management exist:
a) management conducted through administrative and juridical procedures, which partly or wholly require public authorities being involved in recognising and assessing the conflicts and applying measures to solve them;
b) management depending on the disputing parties and based on agreements, arbitration and other similar tools.

Each country, state and region has its own regulations and customs so it is objectively difficult to design a taxonomic framework of the dispute resolution process. Literature on coastal management, especially that of discussing disputes which occurred in the United States after the adoption of the Coastal Zone Management Act (CZMA, 1972), has focused on many case studies relating to the individual countries (Cicin-Sain, Knecht, 1998) but has started providing general views only recently.

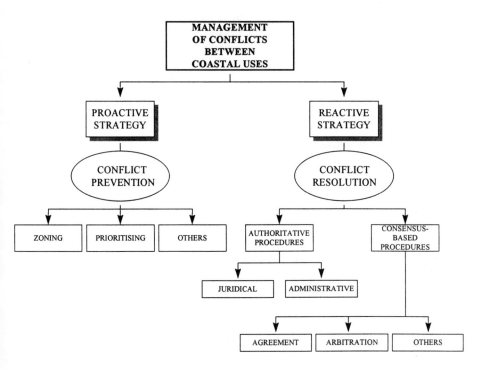

Figure 8.4. Management of conflicts between coastal uses.

8.7.2. PROACTIVE APPROACHES

As coastal management progresses, proactive strategies for conflict management become more sophisticated, and vital in assessing the effectiveness of coastal management can be evaluated. Cicin-Sain (1992: 298–9) notes "Proactive strategies entail an anticipatory approach, typically involve government agencies as major players and may involve private sector mediators as well. A proactive approach would entail, for example, planning efforts for particular marine regions aimed at setting goals for the area (e.g., what should the area look like 5, 10, 20, 50 years from now?; determining carrying capacity for the area; and, deciding on a mix of appropriate uses, both actual and potential)".

These approaches may be applied on both national and subnational scales. In the United States, for example, some patterns of conflict management are being used throughout federal space, and others have been adopted by individual states. Proactive conflict resolutions can be developed in many ways and relevant measures vary from country to country. They include:

a) *Zoning*. The adoption of zoning schemes leads to locating uses in various parts of the coastal area, which may be done in such a way as to prevent conflicts. This approach, which will be dealt with in Chapter 10, belongs to urban and regional planning techniques and needs to be tailored to the relevant specific requirements when applied to coastal areas involved in ICM programmes.

b) *Prioritising*. Coastal management programmes include the list of existing, expected and potential coastal uses ranking them in accordance with the management goals. Where a conflict may materialise, the programme's priorities are applied in order to prevent it.

8.8.   How to Present Conflicts

After the present and previous discussions, one comes to realise how different is to assess the coastal use framework, the use-use and use-ecosystem relationships. The assessment of the conflicting relationships between uses and between users is a particularly hard task. But it is also an unavoidable task, since it leads coastal managers and planners to deal with a principal need of ICM, namely that of minimising impacts from the coastal use structure and optimising the pursuit of sustainable development.

For these reasons, the general approach to conflicts between uses and users may be concluded by presenting a breakdown (Table 8.6) summarising the sequence of subjects which should be handled by coastal managers in order to (i) provide assessment tailored to the ICM principle and the objective of sustainable development, and (ii) design suitable measures for controlling the coastal use structure and implementing its sustainable development.

As regards the literature on conflicting uses, numerous case studies are concerned with the Atlantic waters surrounding the United States and Europe, and the semi-enclosed and enclosed seas facing the south-eastern Asian and western European coasts. Apart from international conflicts, literature only contains a few short accounts of conflicts in the Mediterranean. The reason must not only found in the lack of conflicts, since the use of Mediterranean coastal areas and islands has been characterised by many conflicting situations on both the international scale—where antagonistic economic and political interests are self-evident—and the local scale, where competition between stakeholders and different social views of how the coastal ecosystem, landscape and cultural heritage should be managed, have diffused. Two other reasons may be found. On the one hand, coastal management programmes, triggering social communication and discussion on the future of coastal areas, have not been widely adopted in the Mediterranean, and therefore a gap exists between this framework and that of other regions where this process started ten or twenty years ago. On the other hand, due to cultural and historical factors, in the Mediterranean mechanisms of social communication adapted to discussing conflicts have not arisen. This weak context has not stimulated analysts to focus on coastal conflicts and, consequently, many conflicts are not only latent but also not still assessed. There is no doubt that, in some countries, this situation is going to change because, as a result of three-years preparatory work, the European Commission is adopting guidelines on coastal management.

TABLE 8.6.   A breakdown of views to approach conflicts

Subjects	Objective and nature of assessment	Relevance to sustainable development
Type of use	To assess conflicting uses according to their role in the dynamics of the coastal system. Example: a conflict between recreation and aquaculture.	To provide a useful database on conflicts for the design and implementation of the ICM programme.
Kind of user	To relate conflicting uses to the legal status of users. Example: a conflict between state agency and local authority.	To assess the role of decision makers regarding the design and implementation of the ICM programme.
Legal status of use location	To focus on the legal regime of the coastal space where conflicts have taken place, considering (i) the land areas, (ii) the internal waters, and (iii) the national maritime jurisdictional zones. Example: a conflict between (i) oil and gas industry operating in the claimed continental shelf, and (ii) fishery operating in the overhanging water column, which legally pertains to the high seas.	To know how juridical and jurisdictional frameworks influence the dispute between users and evaluate its relevance to ICM.
Causes of the conflict	To identify and evaluate the causes for the conflicts between uses and users. Example: one use (naval exercises) is organised in terms that do not tolerate another use (merchant navigation).	To acquire useful basic information to resolve conflicts with the final goal of optimising the effectiveness of coastal management and its consistency with the integration principle.
Environmental implications	To describe the consequences that are, actually or potentially, provoked by the conflict on the local ecosystem. Example: one use (seaport breakwater) alters the sedimentation processes changing the scenic values of another use (diving).	To evaluate the impacts on the ecosystem's integrity bearing in mind that its properties (biodiversity, resilience, etc.) are very pertinent to the prospect of carrying out ICM.
Social implications	To describe the consequences that are, actually or potentially, provoked by the conflict on the local economic system and social context. Example: one use increases the labour employment while the other merely improves the quality of life.	To evaluate the consequences of conflicts on the economic efficiency and social equity of the coastal system, including the rights of future generations.
Conflict tractability	To evaluate the conflicts according to the possibility of solving or mitigating them. Example: one user pursues the evolution of the coastal system in terms of implementing the upper tertiary activity, but does not tolerate the growth of heavy coastal industry, such as iron and steel plants.	To assess whether and how public authorities can be efficient in mitigating the intensity of disputes between users. To assess how local communities can participate in this process.
Conflict management	To evaluate possible strategies which may prevent, mitigate or resolve conflicts.	To design principles, regulations and procedures to deal with conflicts according to the integration principle and with the final aim of pursuing the sustainable development of the coastal area.

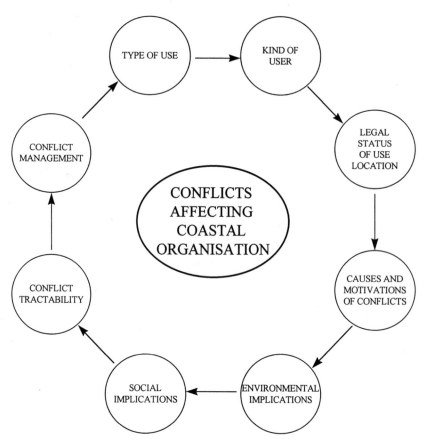

Figure 8.5. Conflicts affecting coastal organisation: subject relevant to integrated coastal management.

8.9.   A Guideline for Managers and Planners

As a conclusion, it may be useful to highlight the distinct roles of the coastal manager and planner, the former being concerned with analysing, monitoring and evaluating the management process, the latter with the design of the management programme's components, sectoral and spatial planning and technical tasks. A decalogue of principles and tasks designing their role regarding the need to approach the conflicts between coastal uses, and between coastal users, may be sketched as follows.

8.9.1. MANAGERS' ROLE

a) *Monitoring conflicts*. A permanent view of the actual and potential conflicts is provided.

b) *Classifying conflicts.* The assessment of conflicts is provided and implemented classifying them in accordance with approaches from both literature and guidelines by inter-governmental organisations.

c) *Designing the possible conflicts brought about by the coastal system dynamics.* Scenarios of possible conflicts depending on the changes on coastal use structure, and coastal area organisation, are built up to help the manager act.

d) *Assessing the social perception of conflicts.* This is done using *had hoc* tools, such as public debates, and public opinion pools, thereby evaluating the social reaction to measures employed to prevent, mitigate and resolve conflicts.

e) *Optimising proactive approaches.* Proactive conflict management is designed consistently with the possibility of benefiting from the collaboration of decision makers, including non-governmental organisations, and social groups.

f) *Disseminating correct and comprehensive information.* To implement social perception of conflicts and to favour agreements between conflicting users, correct and comprehensive information about actual and potential conflicts is provided frequently.

g) *Making the local community aware of the need to prevent or resolve the conflicts referring to sustainable development principle.* The social perception of the need to pursue sustainable development as the final goal of coastal management is implemented with *ad hoc* actions. These focus on the need to prevent or resolve those conflicts which condition this pursuit.

h) *Disseminating correct views of the role of philosophical conflicts.* The role of values, including ethical and existential, is regarded as having the greatest background influence on the effectiveness of reactive and proactive procedures. Suitable actions, such as public discussion, urge the local community to consider the protection of values as an expression of cultural heritage, and as a way of dealing with conflicts.

i) *Favouring conflict resolution through agreements between parties.* Special measures are adopted to privilege agreements, arbitration and the adoption of other similar measures.

j) *Using prioritising methods to prevent and resolve conflicts.* Uses are prioritised according to the end (sustainable development) and contingent goal of ICM, and conflicts are prevented and resolved by prioritising the list of uses.

8.9.2. PLANNER'S TASKS

a) *Taking a holistic view of the use structure.* Planning becomes more effective when the coastal use structure is regarded as the principal means of having a holistic view of the coastal area.

b) *Designing a global view of conflicts.* Moving from the coastal use structure, plans may be carried out by viewing actual and potential conflicts globally. The individual conflict is not considered *per se,* but is related to the whole set of existing and expected conflicts involving the coastal area. The final goal consists of: (a) designing the global stress which the coastal system has undergone, or is expected to undergo, due to conflicts pressurising the coastal organisation, and (b) acquiring a clear idea of the relevance of the individual conflict to stresses.

c) *Focusing on the crucial conflicts*. Special attention is paid to those conflicts which have a deep influence on the evolution of the coastal system.

d) *Designing scenarios on conflicts*. Moving from the existing coastal use structure and bearing in mind the nature and goal of the individual plan, the possible conflict framework, referred to the short-and mid-term, is set up.

e) *Designing the optimum planning approach*. By evaluating the scenarios on possible/expected conflicts brought about by operating the plan, the best approach to conflicts may be designed and incorporated in the plan.

f) *Evaluating the influences of conflicts on the local ecosystem*. The impacts on the local ecosystem from existing conflicts and from the measures designed to prevent and mitigate conflicts included in the individual plan, are assessed by *ad hoc* procedures, which are mostly concerned with the environmental impact assessment (EIA).

g) *Evaluating the influence of conflicts on the human community*. A similar approach is adopted towards impacts on the human community by designing *ad hoc* social impact assessment (SIA) measures.

h) *Designing reactive and proactive procedures*. A set of reactive and proactive procedures is defined and incorporated in the plan paying special attention to devising proactive procedures.

i) *Using the zoning approach to prevent conflicts*. As regards the spatial plans, the coastal area is zoned in such a way as to avoid conflicts.

j) *Setting up a monitoring system*. Where needed, a monitoring apparatus is designed and incorporated in the individual plan, in such a way as to enable the coastal manager to master conflicts efficiently.

The adoption of the above-code of conduct by coastal managers and planners, or any other code tailored to the principle of integrated management, implies that the decision-making system concerned follows these guidelines when organising and carrying out actions. Most of Agenda 21, Chapter 17 regards this as an unavoidable need, and so naturally the role of the decision-makers involved in the coastal area will be dealt with in the next Chapter.

CHAPTER 9

THE COASTAL DECISION-MAKING SYSTEM

This Chapter will present a view of the decision-making system involved in coastal area management, focusing on the role of the decision-makers and proposing a useful methodological approach to integrated management programmes. The following questions will be discussed:

Questions	Figures	Tables
What criteria may be used to assess the role of decision-making systems in coastal management?	9.1	
What criteria may be adopted to assess the links between the legal framework and the decision-making system?	9.2	92, 9.3
How may the integration of the decision-makers be approached?	9.3	
What methods may be applied to focus on the role of the decision-making system?		9.4, 9.5, 9.6
How important are the decision-making processes in setting up integrated management programmes?	9.4	
How may the coastal manager and planner deal with the decision-making system?	9.2	

9.1. The Relevance of the Decision-Making Apparatus

Since the mid-1980s, literature has stressed the importance of legal and institutional arrangements and has regarded them as the main concerns in setting up efficient coastal management. Basically, this is the correct approach because management programmes could not be effective if the legal framework and institutional system did not support an efficient decision-making system. However, the adoption of Agenda 21, Chapter 17, has not merely resulted in planning greater importance on the subject than in the past, but has also posed some crucial questions:
a) how the role of the decision-making system should be defined in accordance with the principle of sustainable development;
b) how the decision-making system should be organised to carry out ICAM programmes efficiently;
c) how the roles of the individual decision-making centres should be assessed in order to evaluate their consistency with the ICM programme;
d) how the legal and institutional framework is to be assessed to evaluate its ability to support decision-making effectively.

9.2. Final and Contingent Objectives

The approach is based on the legal framework and decision-making system being consistent with the coastal management's goals. According to Agenda 21, Chapter 17, two levels of objectives have come to the fore:
* the *higher level*, which is concerned with the final objective, i.e. sustainable development;
* the *lower level*, which is concerned with the specific objectives pursued in the individual area, i.e. the contingent objective of management.

It goes without saying that coastal management is effectively integrated where perfect coherence has been established between the two levels. This is the *teleological equation of ICM*.

The higher level (*final objective*) includes the three components of sustainable development, which have already been presented and discussed (mostly in Chapter 1.3): (i) integrity of the ecosystem, (ii) economic efficiency, (iii) social equity including rights of future generations. The lower level (*contingent objective*) includes goals differing from area to area according to a wide range of factors, such as the nature of the coastal ecosystem, economic organisation, cultural background, social perception of the role of the coastal area, and the geopolitical context. As a consequence, the coastal manager and planner are required to design a specific framework of goals for each area. According to the scale on which the area is considered, this framework changes becoming more or less specific. For example, a general framework arises when all the Mediterranean coastal areas are considered as a whole. However, it becomes more specific when an individual country is considered and even more so when focusing on a given area, such as a coastal French *département*.

Regarding the national scale, Agenda 21, Paragraph 17.6, stated that management "mechanisms could provide, *inter alia*, for:
a) preparation and implementation of land and water use and siting policies;
b) implementation of integrated coastal and marine management and sustainable development plans and programmes at appropriate levels;
c) preparation of coastal profiles identifying critical areas, including eroded zones, physical processes, development patterns, user conflicts and specific priorities for management;
d) prior environmental impact assessment, systematic observation and follow-up of major projects, including the systematic incorporation of results in decision-making;
e) contingency plans for human induced and natural disasters, including likely effects of potential climate change and sea level rise, as well as contingency plans for degradation and pollution of anthropogenic origin, including spills of oil and other materials;
f) improvement of coastal human settlements, especially in housing, drinking water and treatment and disposal of sewage, solid wastes and industrial effluents;

g) periodic assessment of the impacts of external factors and phenomena to ensure that the objectives of integrated management and sustainable development of coastal areas and the marine environment are met;

h) conservation and restoration of altered critical habitats;

i) integration of sectoral programmes on sustainable development for settlements, agriculture, tourism, fishing, ports and industries affecting the coastal area;

j) infrastructure adaptation and alternative employment;

k) human resource development and training;

l) public education, awareness and information programmes;

m) promoting environmentally sound technology and sustainable practices;

n) development and simultaneous implementation of environmental quality criteria".

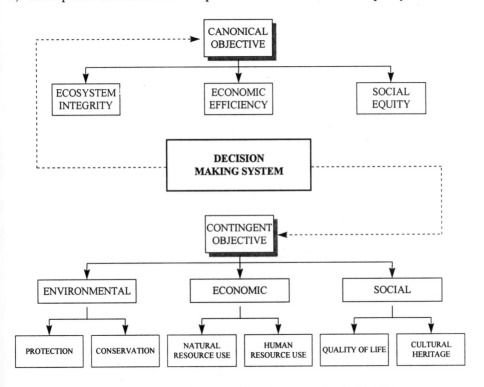

Figure 9.1. Teleological reference bases for the decision-making system involved in ICM.

Evidently, despite referring to actions, this framework is widely concerned with objectives and shows the role of the coastal manager and planner to be intriguing. Shifting to the subnational scale, each coastal area is characterised by its own local goals. Table 9.1 shows, for example, the possible contingent goals of two hypothetical, developed and developing, coastal regions.

TABLE 9.1. Coastal management objectives. Possible frameworks

Developing coastal area (example: a Maghreb coastal area)	Developed coastal area (example: a north-western Mediterranean coastal area)
The objective areas where major differences between developed and developing regions exist are shaded.	
Integrity of ecosystem	
Fresh water management, including the protection of traditional water use techniques	Fresh water management
Aquifer protection	Aquifer protection
Dune conservation	Bush and forest conservation
Protection of endangered species	Protection of endangered species
Protection of the ecosystem against human-induced natural disasters	Protection of the ecosystem against human-induced natural disasters
Protection of the ecosystem against pollution	Protection of the ecosystem against pollution
Economic efficiency	
Protection of man-made structures against sea-level rise	Protection of man-made structures against sea-level rise
Protection of man-made structures against natural disasters	Protection of man-made structures against natural disasters
Development of petrochemical and phosphate industry	Conversion of iron and steel, and petrochemical industry into other coastal uses
Food production (increasing role)	Food production (decreasing role)
Development of hightech manufacturing plants (rarely pursued)	Development of hightech manufacturing plants (intensely and diffusely pursued)
Renewable energy-based plant (not perceived as essential)	Renewable energy-based plant (diffusely perceived as essential)
Building up of technological poles (rarely pursued)	Building up of technological poles (intensely and diffusely pursued)
Seaports: integration of handling facilities with manufacturing development	Seaports: container terminal development
Cruise and passenger seaport development	Cruise and passenger seaport development
Protection of historical settlements against damage from mass tourism	Waterfront revitalisation
Fishery development	Aquaculture development
Recreational use development	Recreational use revitalisation
Social equity (including cultural heritage)	
Implementation of social access to ecosystems	Implementation of social fruition of ecosystems
Implementation of social access to cultural heritage	Implementation of social fruition of the cultural heritage
Establishment of natural parks and protected areas	Establishment of natural parks and protected areas
Protection of archaeological remains	Protection of archaeological remains
Conservation of historical sites	Conservation of historical sites
Implementation of social education	Implementation of social education
Protection and conservation of local folk cultures	Optimisation of social access to maritime museums

9.3.   The Legal Framework

Sustainable coastal development may be pursued only if there is an adequate legal framework to which coastal managers and planners may refer. The role of legal frameworks was tackled indirectly in Chapter 4, where focus centred on national maritime jurisdictional zones and administrative areas. Bearing that approach in mind it is useful to note that the legal framework is characterised by some basic features:
a) it consists of two main components, international and national law;
b) the justification of international law derives from the consensus by states, which is technically expressed by conventions, treaties and agreements of various kinds;
c) the justification of national law derived from the national sovereignty;
d) as a result, international law is applied in the single country only if the state resolves to incorporate it, through ratification, in its own legal system;
e) the juridical rules are based on the principle of hierarchy according to which the rule included in the lower level cannot derogate from the rule included in the higher level, while the latter may modify the former.

As a result, each state is endowed with its own legal framework constituted by:
a) *national law*, consisting of a hierarchical legal structure including the constitution, laws, law-decrees, decrees, orders and others; and
b) that part of the *international law*—i.e., general conventions, multi-national conventions, treaties, protocols and others—which was ratified and, so, was incorporated in the national legal system.

This two-level framework has self-evident influences on the methodological approach to coastal management since it leads to considering two scales distinctly.

The national scale—When national coastal policy is considered, the legal framework to be taken into account includes: (i) the international regulations (general and multi-national conventions, bilateral agreements, and other juridical tools), and (ii) the national regulations adopted by the legislative power (such as the Parliament), and the executive power (such as the President of Republic, the Cabinet Council, and the individual Ministries), and governmental agencies (such as the energy agency or fishery agency). In this framework, a two-level institutional framework may be considered consisting of international organisations and governmental authorities.

The sub-national scale—When coastal management is dealt with on the scale of a state belonging to a federal state, and on a local scale (such as *départments* in France, municipalities, and others), also the local regulations from City Hall, regional and provincial authorities, and others must be taken into account. Consequently, a more complicated framework is to be dealt with by analysts, managers and planners, which could consist of a three-level system: supra-national regulations (ratified by the individual state), national law, and local regulations.

On both national and sub-national scales, regulations and any other juridical tools form a hierarchical system, since the individual regulation claimed at a given level (e.g., adopted by a City Hall) cannot derogate from an upper-level regulation (e.g., a national law). Bearing in mind this principle, the breakdown presented in Table 9.2

can be regarded as a useful starting basis to design the legal framework concerned with the individual coastal areas.

TABLE 9.2. The legal framework of coastal management. A breakdown of its hierarchical setting[*]

Hierarchical setting of the legal system	Features relevant to coastal management	
1st level: *International law*		
Sources of law	Steps of application on the national context	
	Adoption	Ratification
1. International convention		
2. Protocol to the international convention		
3. Regional convention		
4. Protocol to the regional convention		
5. Bilateral treaty		
2nd level: *National law*		
Sources of law	Geographical coverage	
	The national space	A subnational space
1. Constitutional law		
2. National law and its regulations		
3. Law-decree, legislative decree and its regulations		
3rd level: *Subnational regulations*		
Types of regulations	Geographical coverage	
	The whole coastal area	A part of the coastal area
1. Regional law and its regulations		
2. Regulations from local branches of governmental departments		
3. Regulations from local authorities		

[*] The above table presents a logical framework. Where an individual coastal area is considered the right section of the table may be filled in with the data on the local situation.

The Mediterranean may be considered a useful case study to focus on the role of legal frameworks and, particularly, on the issues to be dealt with when efforts to harmonise national policies are focused on. Due to the multitude of political systems influenced by different historical roots and cultural backgrounds, the Mediterranean states are characterised by legal systems differing remarkably from each other, particularly when the northern and southern sides are compared.

Restricting the view to the international law that was incorporated into the national systems, the Mediterranean states can be clustered into two categories:

a) the EU side (at present: Spain, France, Italy, Greece), whose states are subject to international conventions, the Barcelona convention, and regulations from the European Commission;

b) the extra EU side, which is subject to only international conventions and the Barcelona convention.

As an example, Table 9.3 outlines the case of France, where only the international conventions closely concerned with coastal management are mentioned.

TABLE 9.3.   International legal framework concerned with coastal management: the case of France (re: mid-1990s)

Hierarchical organisation of the legal system	State of application of the legal tools	
International conventions	*Adopted*	*Ratified*
1. Law of the Sea Convention, 1982	☐	
2. Convention on Biological diversity, 1992	☐	
3. Convention on Climate Change, 1982	☐	
4. International Convention for Prevention of Pollution from Ships, 1973–1978	☐	☐
Barcelona convention system	*Approved*	*Ratified*
1. Convention for the Protection of the Mediterranean Sea against Pollution, 1976	☐	☐
2. Convention on the Protection of the Marine Environment and the Coastal Region of the Mediterranean (resulting from the amendment of the 1976 Convention), 1995	☐	☐
3. Protocol for the Prevention of Pollution of the Mediterranean Sea by Dumping from Ships and Aircraft, 1976	☐	
4. Protocol concerning Co-operation in Combating Pollution of the Mediterranean Sea by Oil and Other Harmful Substances in cases of Emergency, 1976	☐	
5. Protocol for the Protection of the Mediterranean Sea against Pollution from Land-Based Sources, 1980	☐	
6. Protocol concerning Specially Protected Areas and Biodiversity in the Mediterranean, 1982 (amended in 1995)	☐	
7. Protocol for the Protection of the Mediterranean Sea against Pollution resulting from Exploration and Exploitation of the Sea-Bed and its Subsoil, 1994	☐	☐
8. Protocol on the Prevention of Pollution of the Mediterranean Sea resulting from the Transboundary movements of Hazardous Wastes and their Disposal, 1996	☐	☐
European Union regulations	*Adopted*	*Ratified*
Many regulations on environmental protection, economic development and social life	☐	☐

9.4.   The Role of Decision-Making Systems

Coastal managers and planners are not interested in the legal framework *per se* but are induced to assess it with the aim of framing the role of the decision-making systems into the ICM programme. A conceptual basis, which is useful for understanding the role of these systems' components is provided by the general system theory. Following the approach by Le Moigne (1984: 129–49) the *general model of the coastal decision-making system* could be conceived as operating three roles:

a) *making a choice* between possible development processes and, subsequently, between different coastal management patterns;
b) *designing the objective* of coastal governance on the basis of scenarios on coastal development;
c) *building up strategies* aimed at achieving the goals of coastal development.

The effectiveness of the decision-making system largely depends on the consistency of its organisation with the pursued objectives, on how integration is set up on these three levels, and on the capability of decision-making centres to use *ad hoc* information apparatuses for designing strategies and undertaking actions. Relating this approach to the legal framework, as presented in the previous section, the model of the coastal decision-making system may be broken down into five main components:

a) *inter-governmental organisations*, whose actions are based on international law and, for the marine spaces extending seawards from the baselines, are specifically justified by the law of the sea;
b) *governmental organisations*, including their subnational and local branches, whose actions are based on the national legal system;
c) *private sector centres*, including financial, economic, social, cultural and any other centre not belonging to public administration;
d) *non-governmental organisations* (NGOs), which in their turn can be divided into international NGOs, where they are legally included in rosters of inter-governmental organisations, and national NGOs, where they are connected with governments;
e) *meta-institutional organisations*, consisting of unlawful organisations—such as those concerned with drugs, arms trade, child-slavery and immigrant trade, as well as illegal financial circuits—according to whether they are able to influence the conduct of legal decision-making centres.

Despite being very concise, this view demonstrates that many efforts should be concentrated on the understanding the role of the decision-making systems in coastal management, and on their ability to optimise or frustrate national strategies, local programmes and plans. It could be agreed that this subject needs more intense attention from literature than it has had till now, and particularly it requires taking into account the concrete features and practices of the individual decision-making centres.

In this perspective, there is no doubt that the increasing human pressure on coastal areas—associated with the growing economic interests in many coastal regions and islands—have strengthened the presence and influence of meta-institutional organisations, which require to be considered in depth when the feasibility and effectiveness of management programmes is evaluated. More in general, it is self-evident that, in terms of social impacts, many coastal areas have become very sensitive to the influence of globalisation processes, and therefore the dialectic between global and local systems is going to be increasingly important for management.

This justifies believing that the consideration of the role of decision-making systems may be integrated with that of the social behaviour, particularly the social changes triggered by the advance of post-modern society.

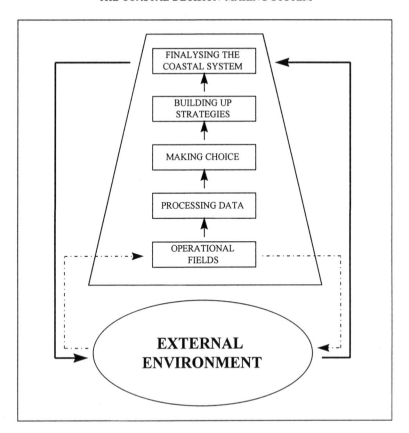

Figure 9.2. The conceptual pattern of the decision-making system according to the theory of the general complexity.

9.5.   Integration of Decision-Making Systems

As is well known, integration is the key word of any sustainable development-aimed policy. *A fortiori*, this key principle is to be referred to coastal management since it deals with one of the most complex parts of the Earth. First, integration refers to the final objective, because the integrity of the ecosystem, economic efficiency and social equity need to be pursued in an integrated way. Then it refers to the decision-making system because all the decision-makers are required to: (i) finalise their organisation, and (ii) co-ordinate their actions to the pursuit of the integrated set of mentioned objectives. It follows that, where ICM programmes are being prepared or implemented, two basic questions have to be dealt with:
a) a *teleological question*: whether and to what extent the organisation of the existing decision-making system, considered as a whole, is consistent with the pursuit of

sustainable coastal development (final goal) and the goals designed by the pro-gramme (contingent goals);
b) an *operational question*: whether the actions of decision-making centres are cor-rectly finalised to the pursuit of goals.

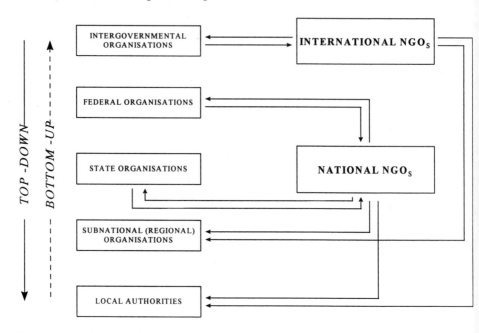

Figure 9.3. Vertical interaction between decision-making systems involved in ICM.

Both questions are emphasised by Agenda 21, particularly Paragraphs 17.116 and 17.119, which should be taken into consideration not only from coastal managers and planners, but also by all the decision-making centres involved in coastal affairs.

The existing organisation of an individual coastal area, as well as that of the set of coastal areas belonging to a country or a state, may be valued taking two basic kinds of integration into account:
a) *the vertical integration*, existing between different levels of the decision-makers concerned with the same sector—for instance, between a governmental department and a local authority on fishing;
b) *the horizontal integration*, involving the decision-makers operating in various sec-tors of the same level—for instance, between two or more departments (energy, environment, public works) of an individual municipality involved in evaluating the environmental impact from coastal thermoelectric plants.

Agenda 21, Paragraph 17.116

"States commit themselves, in accordance with their policies, priorities and resources, to promote institutional arrangements necessary to support the implementation of the programme areas in this Chapter. To this end, it is necessary, as appropriate, to:

i.   Integrate relevant sectoral activities addressing environment and development in marine and coastal areas at national, subregional, regional and global levels, as appropriate;
ii. Promote effective information exchange and, where appropriate, institutional linkages between bilateral and multilateral national, regional, subregional and interregional institutions dealing with environment and development in marine and coastal areas;
iii. Promote within the United Nations system, regular intergovernmental review and consideration of environment and development issues with respect to marine and coastal areas;
iv. Promote the effective operation of co-ordinating mechanisms for the components of the United Nations system dealing with issues of environment and development in marine and coastal areas, as well as links with relevant international development bodies".

Agenda 21, Paragraph 17.119

"States should consider, as appropriate:

i.   Strengthening, and extending where necessary, intergovernmental regional cooperation, the Regional Seas Programmes of UNEP, regional and subregional fisheries organizations and regional commissions;
ii. Introduce, where necessary, coordination among relevant United Nations and other multilateral organizations at the subregional and regional levels, including consideration of co-location of their staff;
iii. Arrange for periodic intraregional consultations;
iv. Facilitate access to and use of expertise and technology through relevant national bodies to subregional and regional centres and networks, such as the Regional Centres for Marine Technology".

9.5.1. VERTICAL INTEGRATION

As can be deduced from the mentioned Paragraphs of Agenda 21, coastal management requires a wide spectrum of vertical integration, since it encompasses numerous levels of decision-makers—from the international to local level. To undertake ICM in an individual coastal area, vertical co-ordination between decision-making centres involves the following levels (fisheries is the sector considered by way of example):

a) *inter-governmental organisations*, e.g., FAO for fisheries and IMO for security in coastal navigation;
b) *governmental organisations*, e.g., Ministry of Agriculture for fisheries and Ministry of Transportation for coastal navigation;
c) *regional authorities*, e.g. the *départments* of a French region operating in fisheries and navigation;
d) *local authorities*, e.g. the municipalities included in the individual coastal area where fishing issues have come to the fore;
e) *economic associations*, e.g. the national fishery association and local fishermen's associations;
f) *non-governmental organisations* involved or interested in living resource exploitation.

The correct application of the integration concept would require co-operation between different levels of decision-making centres developing through efficient top-down and bottom-up relationships. Coastal management guidelines by UN organisations and programmes, as well as those by national governments, typically represent the top-down relationships. On the other hand, the daily experience of the local decision makers and groups brings about inputs towards the top thereby improving the efficiency of the whole decision-making system. Hence, if both tracks are implemented giving shape to feed-backs, then the organisation of the coastal area will become more efficient.

In this regard, Agenda 21, Paragraph 37.2, stresses the importance of implementing capacity building with these statements:

Agenda 21, Paragraph 37.2

"Building endogenous capacity to implement Agenda 21 will require the efforts of the countries themselves in partnership with relevant United Nations organizations, as well as with developed countries. The international community at the national, subregional and regional levels, municipalities, non-governmental organizations, universities and research centres, and business and other private institutions and organizations could also assist in these efforts. It is essential for individual countries to identify priorities and determine the means for building the capacity and capability to implement Agenda 21, taking into account their environmental and economic needs. Skills, knowledge and technical know-how at the individual and institutional levels are necessary for institution-building, policy analysis and development management, including the assessment of alternative courses of action with a view to enhancing access to and transfer of technology and promoting economic development. Technical cooperation, including that related to technology transfer and know-how, encompasses the whole range of activities to develop or strengthen individual and group capacities and capabilities. It should serve the purpose of long-term capacity-building and needs to be managed and coordinated by the countries themselves. Technical cooperation, including that related to technology transfer and know-how, is effective only when it is derived from and related to a country's own strategies and priorities on environment and development and when development agencies and Governments define improved and consistent policies and procedures to support this process".

9.5.2. HORIZONTAL INTEGRATION

Since the coastal area has to be managed as a system and coastal management has to be carried out holistically, all the decision-making centres operating at the same level therefore need to be co-ordinated as closely as possible. For instance, the departments of the individual coastal municipality (such as the departments of economic development, environment, public health, and culture) should be organised in such a way as to operate as one decision-making centre: their operational fields should not overlap, their organisation should be finalised to the same spectrum of objectives, and their actions should be harmonised. Where ICM is undertaken, integration needs to be optimised to the point that the local community perceives the actions by various decision-making centres as by an individual system.

During the meetings of the UNCED Preparatory Committee discussion largely concentrated on the potential difficulties of pursuing both vertical and horizontal organisational integration. Governmental delegations and experts were correctly aware that the main issue would involve integrating the decision-making centres operating in the development and environmental sectors because the sustainable development principle is based on harmonising human actions regarding the ecosystem and economic organisation. However, integration is also needed between other sectors, such as those concerned with cultural heritage. Strong efforts should be made to meet these needs as it usually becomes stressing where decision-making centres are liable to re-organisation and re-orientation processes. In all cases, political interest, sustaining the individual centres, makes this operation rather hard and laborious. The UNCED Pre-paratory Committee found the best way to solve these issues by establishing Commis-sions on Sustainable Development (CSD) at various levels. It was decided that this body would be set up at the highest level—hence the foundation of the UN Commis-sion on Sustainable Development (CSD)—and at regional (multi-national), national and subnational levels.

The Mediterranean gives good illustration of the role of CSD at the regional level. In 1995, (Barcelona, Spain, June 5–10) the Ninth Ordinary Meeting of the Contract-ing Parties and the Meeting of the Plenipotentiaries to the Barcelona Convention re-solved to establish the Mediterranean Commission on Sustainable Development (MCSD). Its prerogatives were determined by the Extraordinary Meeting of the Con-tracting Parties to the Barcelona Convention (Montpellier, France, 1–4 July, 1996). According to that resolution, "the functions of the Commission shall be:

a) to assist in the formulation and implementation of a regional strategy of sustainable development in the Mediterranean (...);

b) to consider and review information provided by the Contracting Parties (...);

c) to review at regular intervals the cooperation of MAP with the World Bank and other international financial institutions, as well as the European Union, and to ex-plore ways and means for the strengthening of such cooperation (...);

d) to consider information regarding the progress made in the implementation of relevant environmental conventions (...);

e) to identify technologies and knowledge of an innovative nature for sustainable de-velopment in the Mediterranean region and to provide advice on the various means for their most effective use (...);

f) to provide reports and appropriate recommendations to the meetings of the Con-tracting Parties (...);

g) to undertake a four-year strategic assessment and evaluation of the implementation by the Contracting Parties of Agenda 21 (...)".

Commissions on Sustainable Development were also established on the national scale. As regards the Mediterranean, the French *Commission du Développement Du-rable* should be noted.

On the subnational scale, the concrete issues concerned with the individual ICM programme arise. In these cases, horizontal organisational integration has to be pur-sued in the framework of the intermediate level between the national and local scales (e.g., in the framework of an individual province, region, *département*), or on the lo-cal scale, such as in the municipal area. On these scales, all the public administration departments concerned with coastal management should be united in such a way as to create a body wholly devoted to sustainable development. For example, the establish-ment of the "Department of ICAM" within the individual province, or municipality, might be the most effective institutional approach. However, where this is very diffi-cult to achieve a compromise may be proposed. Two of them are worthy of consideration, consisting in the establishment of (i) the local Commission on ICAM, or a Commission on Coastal Sustainable Development, or (ii) a co-ordinating body, e.g. a consultative inter-departmental committee on coastal management.

9.6.   Matrix-Based Assessment of the Decision-Making System

To implement co-ordination between decision-making centres acting at different levels (vertical organisational integration), or at the same level of different sectors (horizontal organisational integration), coastal managers and planners are required to face the need: (i) to identify and assess the decision-making system concerned with the coastal area, and (ii) to prefigure whether and how the system can be implemented. It would be useful to use matrices, of which three are presented below representing the role of decision-making centres facing (i) coastal uses, (ii) the operational areas of coastal management, and (iii) the geographical areas over which cognisance extends.

9.6.1. DECISION-MAKING SYSTEM/COASTAL USE MATRIX

To sketch a comprehensive view of the role of decision-making systems it is useful to set up a rectangular matrix where:
a) *rows* represent the decision-making centres concerned with the coastal area under investigation and classified according to their operational levels and to an a top-down approach;
b) *columns* specify the coastal uses.

The coastal use framework discussed in Chapter 7.4 is categorised in columns according to the need of the manager. If a general view is necessary, only the categories of uses can be included; whereas a detailed view is required, sub-categories and other more specific classification levels can be presented. For instance, if living resource exploitation is to be assessed in detail, the classification shown in Table 7.5 may be adopted. Since the decision-making centres shown in rows vary from area to area, a general approach cannot be given. The Table is hierarchically structured starting with inter-governmental organisations and followed by national organisations and subnational centres.

An example of a *decision-making system/coastal use* matrix is illustrated in Table 9.4, referring to a hypothetical coastal region of the Mediterranean-EU side. For practical reasons, six categories of uses are shown in columns, forming the initial part of the *coastal use framework* presented in Table 7.4.

Having shown the above breakdown, it is now necessary to assess the role played by the decision-making centres concerned with the management of the individual coastal area using a matrix where:
• *rows* represent the framework of decision-making centres;
• *columns* specify their functions.
Functions may be classified according to (i) the criteria followed by managers and planners, and (ii) the concrete features of the coastal area. Table 9.5 shows a useful

classification embracing six functions: planning, management, research, surveillance/control, assistance, promotion.

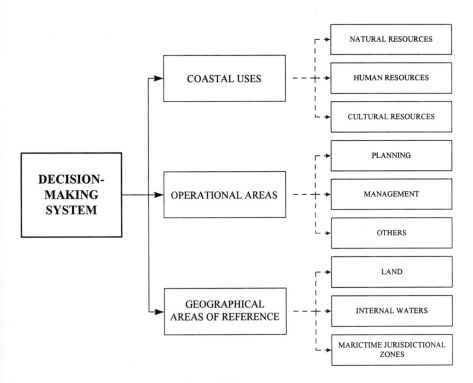

Figure 9.4. The decision-making system facing ICM.

9.6.2. DECISION-MAKING SYSTEM/GEOGRAPHICAL AREAS MATRIX

Chapter 4 discussed how the coastal area consists of different parts, each subject to a legal and jurisdictional regime. Now it would be useful to assess where the individual decision-making centres operate, and how far their stewardship extends. Table 9.6 shows a matrix where:

a) *rows* represent the framework of decision-making centres presented in Table 9.4;

b) *columns* specify their relevant operational geographical areas.

Obviously, other matrices can be conceived and applied, but it would not be appropriate to elaborate any further. Instead, it is essential to note that the matrix-based approach may open the way to processing data by GIS and to implementing the methods concerned with Decision Support Systems (DSSs) analysis, which is a well known chapter of GIS application. In this way, an alternative organisation of the decision-making system can be simulated.

TABLE 9.4.   Decision-making system/coastal use matrix. Re: Mediterranean-EU side

Levels of the decision-making system	Categories of coastal uses					
	1	2	3	4	5	6
Inter-governmental organisations [1]						
UNEP/MAP Co-ordinating Unit	▢	▢	▢	▢		▢
MAP/ Regional Activity Centres	▢		▢	▢		
FAO	▢	▢		▢		
UNESCO/IOC	▢					
IMO						▢
WMO						
IAEA					▢	
World Bank	▢	▢	▢			
National organisations [2]						
National Commission on Sustainable Development	▢	▢	▢	▢		▢
Ministry of Environment	▢		▢	▢	▢	▢
Ministry of Agriculture		▢		▢		
Ministry of Industry					▢	▢
Ministry of Public Works					▢	▢
Ministry of Transportation					▢	▢
Ministry of Human Health			▢		▢	▢
Ministry of Tourism						
Manufacturers' Association					▢	▢
Seaport and Transport Associations					▢	
Navigation Companies Associations						
Landowners Associations		▢		▢		
Trade Unions		▢			▢	▢
Sub-national decision-making centres						
Regional authority (Départements, Regioni and others)	▢	▢		▢		
Province	▢	▢		▢		
Municipalities	▢	▢		▢		
Coast Guard						▢
Ministry of Environment, local branch	▢	▢		▢	▢	▢
Ministry of Agriculture, local branch		▢		▢		
Ministry of Industry, local branch					▢	
Ministry of public works, local branch					▢	▢
Ministry of Transportation, local branch					▢	▢
Ministry of Human Health, local branch				▢	▢	▢
Ministry of Tourism, local branch						

Codes:
1. Ecosystem conservation
2. Agriculture and livestock
3. Natural vegetation use
4. Living resource exploitation
5. Energy production
6. Mineral resource exploitation (land and marine fields).

Notes:
[1]   At this level international NGOs operate in co-operation with inter-governmental organisations.
[2]   At this level national NGOs operate in co-operation with governmental organisations.

9.6.3. DECISION-MAKING SYSTEM/OPERATIONAL AREAS MATRIX

TABLE 9.5.   Decision-making system/operational areas matrix. Re: Mediterranean-EU side

Levels of the decision-making system	Operational areas					
	1	2	3	4	5	6
Inter-governmental Organisations [1]						
UNEP/MAP Co-ordinating Unit			☐		☐	☐
MAP: Regional Activity Centres			☐		☐	☐
FAO			☐		☐	☐
UNESCO/IOC	☐		☐		☐	☐
IMO	☐		☐	☐	☐	☐
WMO			☐		☐	☐
IAEA			☐		☐	☐
World Bank						☐
National Organisations [2]						
National Commission on Sustainable Development	☐					☐
Ministry of Environment	☐	☐	☐	☐	☐	☐
Ministry of Agriculture	☐	☐		☐		
Ministry of Industry	☐	☐		☐		
Ministry of Public Works	☐	☐		☐		
Ministry of Transportation	☐	☐		☐		
Ministry of Human Health	☐	☐		☐		
Ministry of Tourism	☐	☐		☐		
Manufacturers' Association		☐				☐
Seaport and Transport Associations		☐				☐
Navigation Companies Associations		☐				☐
Landowners Associations		☐				☐
Trade Unions		☐				☐
Sub-national decision-making centres						
Regional authority (Départments, Regioni and others)	☐	☐		☐		
Province	☐	☐		☐		
Municipalities	☐	☐		☐		
Coast Guard				☐	☐	
Ministry of Environment, local branch	☐	☐		☐		
Ministry of Agriculture, local branch	☐	☐		☐		
Ministry of Industry, local branch	☐	☐		☐		
Ministry of public works, local branch	☐	☐		☐		
Ministry of Transportation, local branch	☐	☐		☐		
Ministry of Human Health, local branch	☐	☐		☐		
Ministry of Tourism, local branch	☐	☐		☐		
University and research bodies		☐	☐	☐		

Codes: 1. Planning, 2. Management, 3. Research, 4. Surveillance and control, 5. Assistance, 6. Promotion.

Notes:

[1]   At this level international NGOs operate in co-operation with inter-governmental organisations.

[2]   At this level national NGOs operate in co-operation with governmental organisations.

TABLE 9.6. Decision-making system/geographical areas matrix. Re: Mediterranean-EU side

Levels of the decision-making system	Geographical areas					
	1	2	3	4	5	6
Inter-governmental Organisations [1]						
UNEP/MAP Co-ordinating Unit			☐	☐	☐	☐
MAP: Regional Activity Centres			☐	☐	☐	☐
FAO			☐	☐	☐	☐
UNESCO/IOC			☐	☐	☐	☐
IMO			☐	☐	☐	☐
WMO						
IAEA						
World Bank						
National organisations [2]						
National Commission on Sustainable Development	☐	☐	☐	☐	☐	☐
Ministry of Environment	☐	☐	☐			☐
Ministry of Agriculture [4]	☐	☐				☐
Ministry of Industry	☐	☐			☐	
Ministry of Public Works	☐	☐				
Ministry of Transportation	☐	☐				
Ministry of Human Health	☐	☐				
Ministry of Tourism	☐	☐	☐			
Manufacturers' Association	☐	☐				
Seaport and Transport Associations	☐	☐				
Navigation Companies Associations	☐	☐	☐			
Landowners Associations						
Trade Unions						
Sub-national decision-making centres						
Regional authority (Départments, Regioni and others)	☐	☐				
Province	☐	☐				
Municipalities	☐	☐				
Coast Guard	☐	☐	☐	☐		
Ministry of Environment, local branch	☐	☐	☐			☐
Ministry of Agriculture, local branch	☐	☐				☐
Ministry of Industry, local branch	☐	☐				
Ministry of public works, local branch	☐	☐				
Ministry of Transportation, local branch	☐	☐				
Ministry of Human Health, local branch	☐	☐				
Ministry of Tourism, local branch	☐	☐	☐			
Fishermen Associations	☐	☐	☐			☐

*Codes:*1. Land, 2. internal waters, 3. territorial sea, 4. contiguous zone, 5. continental shelf, 6 exclusive fishery zone

Notes:

[1] At this level international NGOs operate in co-operation with inter-governmental organisations.

[2] At this level national NGOs operate in co-operation with governmental organisations.

[3] Including fisheries and aquaculture.

9.7.   Decision-Making Processes

To approach coastal area management with close reference to the guidelines provided by Agenda 21, the decision-making processes are of prime interest. Management is more likely to be successful when decision-making processes and their consistency with ICM are assessed. Matrices on the decision-making system organisation are to some extent introductory steps to understanding how existing decision processes may be assessed and new patterns designed. In particular, the "decision-making system/operational areas" matrix leads the coastal manager to assess how many centres are involved in the individual procedure (e.g., fishing permits, establishment of a protected area) and how complicated the administrative way ahead is. To date, literature has not provided well defined methodological criteria to tackle this subject area, but, nevertheless, some general considerations can be made.

First, the approach to decision-making processes is expected to pursue two objectives:
a) the assessment of the procedure and its relevance to, and consistency with, the ICM concept;
b) the optimisation of processes, making them as consistent as possible with the integration principle.

Consistency of the decision-making processes with the integration principle requires the process being conducted according to the guidelines provided by Agenda 21 and subsequent materials from the UN system and other inter-governmental organisations. In particular, two principles are relevant. First, the *capacity building* principle, in which decision-making processes should be oriented to maximise the local capacity for implementing sustainable development-inspired approaches to the coastal area. For example, the decision-making processes with regard to the establishment of a marine museum should aim to include materials and interactive tools capable of stimulating research into marine ecosystems and social education on protecting the ecosystem's integrity. Secondly, the *social participation* principle, in which each decision-making process should maximise the participation of local groups. For example, the decision processes leading to the permit for the installation of an aquaculture plant should leave room for fishermen and other interested social groups to participate in the plant to ensure that coexistence and harmony prevails between the conventional activity, i.e. fishing, and the new one.

Having dealt with the consistency of decision-making processes, it is now necessary to consider their effectiveness. At least two principles, both widely discussed in literature, need to be recalled. First, the *precautionary principle*, in which coastal development is more effective when the decision-making process prevents the coastal system from such damage, as human-induced natural disasters. Secondly, the *proactive principle*, in which coastal organisation is more effective when the decision-making process prevents conflicts between coastal users.

The efficiency of decision-making processes should be monitored as well. Among the numerous criteria and parameters used to assess and measure this property of coastal management, two may be considered useful. First, the time elapsing between the adoption of the decision and its initial operational consequences, i.e. between the initial and final steps of the decision-making process: the shorter the time means greater efficiency. Secondly, the number of administrative and extra-administrative bodies participating in the process: the fewer the bodies means greater efficiency.

9.8.   A Tentative Agenda for Managers and Planners

During the take off (1970s) and drive to maturity (1980s) stages of coastal management only some of the literature dealt with the role of decision-making systems and processes which, however, was discussed only in relation to the legal and institutional frameworks, and merely on the national scale. As a result, only the legal and institutional frameworks regarding the existing or possible national approaches to coastal area management were taken into account. Where the guidelines from Agenda 21 are assumed as reference bases and attention is focused on the local scale, the approach needs to change. The major change is due to the fact that the legal and institutional frameworks must be considered in view of designing mechanisms to enable the ICM, regarding the individual coastal area, to be undertaken and efficiently implemented.

Consequently, when the coastal manager and planner are led to put new subjects in their agenda, and to optimise their role, they could, as a preliminary approach, aim at:
a) assessing how the legal framework supports and influences the role of the decision-making processes involved in coastal management;
b) assessing the existing decision-making system with reference to no less than three variables: the coastal uses with which each decision-making centre is concerned; the operational area by which it is characterised; and the geographical area in which it operates;
c) evaluating the vertical and horizontal integration of decision-makers;
d) evaluating the consistency of the system with the ICM principle;
e) including the data referring the mentioned subjects in a GIS apparatus with the aim of setting up a computer-based Decision Supporting System (DSS);
f) simulating possible adjustments to the decision-making system in order to optimise its role in carrying out ICM programmes;
g) monitoring the effectiveness of the decision-making system with the aim of evaluating, step-by-step, how it evolves *vis à vis* the pursuit of coastal sustainable development.

The extent of this agenda demonstrates that the decision-making system and processes will be probably one of the most discussed subjects in literature and will be perceived as increasingly relevant to ICM programmes.

CHAPTER 10

INTEGRATED COASTAL MANAGEMENT PROGRAMMES

As a conclusion of the book, a view of the Integrated Coastal Area Management (ICAM) process will be presented. The subject areas discussed in previous Chapters will be framed in the structure of ICAM programmes. A tentative design of the ICAM programme will be provided and the core issues to be tackled will be emphasised. The following questions will be discussed:

Questions	Figures	Tables
What is the ICAM process?	10.1	10.1, 10.2
How may the ICAM process be justified and decision-making systems induced to undertake it?	10.2	10.3, 10.4
How may the ICAM process be initiated?	10.3 to 10.6	
What are the contents and development stages of the ICAM programme?	10.3 to 10.6	10.5 to 10.9
Which plans should be included in the ICAM programme?	10.3 to 10.6	10.10, 10.11
How may the ICAM programme be implemented?	10.3 to 10.6	10.12
What role does monitoring play in the ICAM programme implementation?		10.13
How may the ICAM process be evaluated?		10.14

10.1. Mises au Point

As a final step, this Chapter tackles the Integrated Coastal Area Management (ICAM) programme but, due to its large scope, merely presents a range of fundamental aspects. Most literature on ICM is concerned with the national scale, and frequently discusses such questions as how *national strategy* pursuing ICM may be triggered, adopted and enforced, what institutional frameworks may be assumed as reference bases, and what political, legal and economic mechanisms may be put in motion. These approaches usually leave the local scale in the background, or are not concerned with it. As a result, the question of how coastal management programmes may be initiated in the *individual area* is considered less or not at all. Also guidelines from inter-governmental organisations may be divided into two groups. On the one hand, those approaches, such as the contributions from FAO (1992) and OECD (1993a), which are concerned with the national scale and, on the other hand, those very few approaches, such as that from UNEP (1995), which are also fairly concerned with the local scale.

This Chapter will only deal with the *local scale*, presenting and discussing how ICM may be undertaken in the context of the administrative local framework, such as provinces, *départments*, and regions. There are basically four reference areas: managers (scientific) and planners (technical), decision-making centres (political) and local community (social).

In both scientific literature and materials from inter-governmental organisations there is no consensus on the meaning of some basic terms, such as "programme", "plan", "strategy plan", "master plan", and others. To avoid misunderstanding, these terms will be adopted in this Chapter:

- *Integrated Coastal Area Management* (ICAM) will be intended as a process essentially aimed at pursuing the sustainable development of the coastal area through close integration of development goals and ecosystem protection.
- The *ICAM programme* will be intended as the time sequence of actions designed to undertake and implement integrated management.
- *Planning* will be meant as the set of patterns through which coastal resources—natural, human and cultural—are exploited and protected. Planning does not coincide with the programme but is a part of this. Its content is essentially technical.
- *Management tools* are intended as all the instruments—legal, economic, financial, fiscal, cultural and communication—through which the programme is implemented, monitored and evaluated.

10.2. The ICAM Process

The view of ICAM being regarded as a mid- or long-term process is emphasised by both scientific literature (Cicin-Sain and Knecht 1998: 57–60) and materials from the UN system and other inter-governmental organisations, such as the EU. As regards the UN system, it is worth considering the approach from the World Bank (1993) and UNEP (1995).

According to the World Bank, the ICAM process is started by triggering the need for ICAM and evolves by designing, operating and implementing the programme. Basically, but not wholly, this view is reflected in the UNEP approach, which designs coastal management as a process including initiation, planning and implementation. The comparison between these two approaches is presented in Table 10.1. As can be seen, they are conceptually and methodologically supported by relevant experience. The World Bank's view is sensitive to the experience in the framework of programmes funded by the UN system and to the approaches developed in some national contexts, especially the United States. The UNEP view is basically motivated by the experience carried out in the framework of its Ocean and Coastal Areas, Programme Activity Centre (OCA/PAC), and pays special consideration to the coastal management programmes carried out in the Mediterranean by the Priority Actions Programme, Regional Activity Centre (PAP/RAC), which were presented in Table 1.8.

TABLE 10.1.   The ICAM process according to The World Bank and UNEP

The World Bank 1993	UNEP 1995
Stages and phases	Stages and phases
	1st stage
Initiating the effort	Preparation activities
Formulating the plan	Analysis and forecast
	2nd stage
Formal adoption of the ICAM programme	Definition of goals and strategies
Operational phase	Integration of detailed plans
	3rd stage
Programme implementation	Implementation of plans
	Monitoring and evaluation
	4th stage
Monitoring, evaluation and enforcement	

The approach presented in the following Sections aims at implementing these views by jointly taking into account: (i) the global change concept as was formulated in the framework of the international research programmes, and presented in Chapter 2; (ii) the sustainable development concept, as it was proposed in the framework of discussions developed in the course of the 1980s and adopted by the Rio Conference, i.e. including ecological protection, economic efficiency, and social equity; and (iii) some conceptual and methodological contributions inspired by the theory of complexity, which have been the basis of this book, and which were specifically dealt with in Chapters 5 to 7. In practice, the approach to the conceptual and methodological bases of integrated coastal management, developed in previous chapters, will be reflected in the stages and phases marking the ICAM process. In addition, following the literature in the field, integrated coastal management will be regarded as consisting of a long-term sequence of scientific, political and social actions. The perspective from which this process may be conceived is shown in Figure 10.1 and Table 10.2, where the ICAM programme is presented as consisting of six stages: (i) justification, (ii) initiation, (iii) preparation, (iv) planning, (v) implementation, and (vi) monitoring and evaluation.

This is merely one possible view of the ICAM process, and literature has provided others, some of which have played important roles in influencing both national policies and the design of programmes on the local scale. In this respect, two examples may be considered. First, the well known Sorensen and McCreary's model (1990: 32) merits being taken into consideration, since it has been used as a basis for many coastal management programmes, and was substantially adopted by UNEP (1995: 22). It was presented in the years when the preparation of the Rio Conference was operated, and consequently it has drawn attention from scientists and policy-makers involved in the discussion of how the coastal concern had to be dealt with by Agenda 21, that was in preparation. Essentially, this approach is sensitive to coastal policy carried out in the United States during the 1970s and 1980s but, at the same time, it is keen to consider the experiences of other parts of the world. As can be seen from

Figure 10.3, it refers to the national scale and hypothesises that the coastal management process passes through eight stages: (i) incipient awareness, (ii) growing awareness, (iii) national study, (iv) programme creation, (v) programme development, (vi) programme implementation, (vii) programme evaluation, and (viii) rejection or implementation of the programme. Secondly, the pattern provided by Chua (1992: 25) may be regarded as one of the most influential. It is presented under a matrix format. As can be seen from Figure 10.6, the ICAM processes is conceived as consisting of eight stages: (i) triggering stage, (ii) research and analysis, (iii) formulation of plan, (iv) decision making, (v) adoption, (vi) execution, (vii) monitoring and evaluation, (viii) integration.

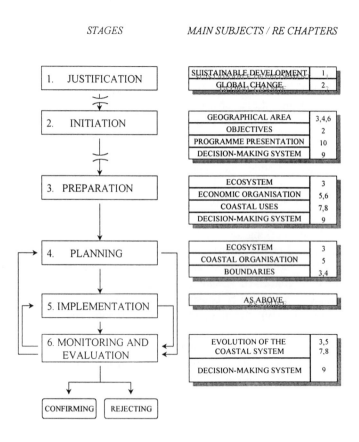

Figure 10.1. The ICAM process, as presented in this Chapter, and the relevant main subjects, which were discussed in previous chapters (indicated in the right side).

TABLE 10.2.   The ICAM process. A basic breakdown

Stages	Basic outputs		
	Scientific	Political	Social
I. *Justification* Objective: to catalyse consensus on the need for ICAM	Identifying the main issues and prospects of coastal protection and development. Proposing a range of environmental and developmental objectives. Proposing the geographical coverage of the ICAM programme.	Encouraging the proclivity of decision-making centres to develop coastal management based on the integration principle. Persuading decision-makers of the political advantages of the ICAM programme.	Diffusing the concept of integrated coastal management. Triggering public awareness of the need for the ICAM programme. Diffusing the conviction that the ICAM programme is the main tool to harmonise short- and long-term prospects.
II. *Initiation* Objective: to undertake the ICAM process	Designing the geographical extent of the coastal area. Defining the objectives of the ICAM programme. Determining the stages of the ICAM programme. Identifying the funding mechanisms for the ICAM programme.	Accepting the proposal for the ICAM programme. Establishment of a co-operation network of local organisations and institutions to undertake ICM. Funding the preparation of the ICAM programme.	Establishing information systems and mechanisms of social participation in the ICAM process. Convening discussion on the specific objectives and content of the ICAM programme. Attracting the co-operation from non-governmental organisations.
III. *Preparation* Objective: to adopt the ICAM programme	Preparation of the ICAM programme, designing strategies and including the master plan. *Id.*, of the main technical documents, including those on funding.	Adoption of the ICAM programme. Establishment of a Steering Committee to carry out the programme.	Presentation of the programme to the local community. Establishing and operating communication and social participation mechanisms (bottom-up principle).
IV. *Planning* Objective: to adopt the ICAM plans and strategies	Setting up the sectoral plans concerned with the leading components of the coastal organisation. *Id.*, the special plans concerned with special purposes (e.g., parks and reserves) of ICAM. *Id.*, mechanisms of data collecting and processing.	Adopting the sectoral and special plans, as well as other actions aimed at ICAM implementation.	Convening consultations with the local community. Catalysing social consensus on ICAM planning and strategies. Implementing tools for social information and participation.
V. *Implementation* Objective: to adopt appropriate plans and measures for the ICAM process	Optimising sectoral and special plans and management tools, including regulatory economic instruments. Developing assessment of environmental impacts. *Id.*, social impact assessment.	Taking actions aimed at implementing the ICAM process.	Implementing social information and participation in the ICAM process.
VI. *Monitoring and evaluation* Objective: to evaluate the outputs of the ICAM process	Re-defining cross sectoral issues. Evaluating the results of the ICAM process. *Id.*, efficiency of management tools. *Id.*, environmental and social impacts.	Confirming or cancelling the ICAM programme. Confirming or cancelling the master plan. Confirming or cancelling individual sectoral and special plans. Enforcing regulatory and economic tools.	Discussing the outcomes of, and impacts from, the ICAM programme. Testing the social propensity to maintain, re-design or annul the ICAM programme.

STAGE I—JUSTIFICATION

The ICAM process may be initiated by catalysing social consensus about the need to move from the conventional sector-based approach to an innovative, holistic approach, aimed at guaranteeing the integrity of the ecosystem and social equity, together with the pursuit of economic efficiency. Besides, decision-making centres should be stimulated and encouraged to undertake integrated management. This pre take off stage is usually called "triggering stage" in literature (The World Bank 1993; Cicin-Sain and Knecht: 41-3) and is to be kept distinct from the initiation stage because it does not seek the decision makers's approval of a proposal on ICAM programme, but just creates the relevant political and social conditions to reach this goal. If the triggering action is successful, the initiation phase begins, where the documents justifying and supporting the ICAM programme are done.

The triggering task acquires different features, and needs different approaches according to the existence or absence of national policy on ICAM. Where national laws and guidelines exist, such as in France through the law on coastal conservancy (*Loi sur le Conservatoire du Littoral*), public opinion makers are in the position to persuade the local decision-making centres and communities of the need to follow the national government guidelines. Clearly, where the adoption of the ICAM programme enables the local decision-making system to get governmental funds, this task is quite easy to carry out. On the contrary, where national policy is not adopted, it is difficult to trigger social awareness of the need for the ICAM programme and stimulate political proclivity to undertake such an initiative. In this case, ICAM has to be justified to persuade the local decision-making system to adopt the programme where necessary for dealing with unavoidable issues or solidifying attractive development prospects.

The more scientists and public opinion makers are successful in demonstrating that, at the present state of the art, integrated coastal management is the best course to harmonise environmental and developmental issues and goals, and to combine short- and long-term strategies, the more decision-makers feel inclined to react positively to the triggering action.

Usually general statements, focusing abstractly on the usefulness of adopting and operating the concept of integration, are unsuccessful. What is needed is the social perception of the daily issues which can be solved, or mitigated, when an ICAM programme is adopted. It may be useful, like other authors have tried, to list the triggers of ICAM. A view of the contributions on these subjects led Sorensen and McCreary (1990: 38-43) to identifying four groups of triggering factors. Cicin-Sain and Knecht (1998: 41-43) provided a breakdown of the triggering factors according to the States. Frequently, the impacts from the individual coastal activities, coastal hazards, and the need to develop promising economic sectors have a pivotal role. A view deduced by Sorensen and McCreary (1990:38-41) is presented in Table 10.3

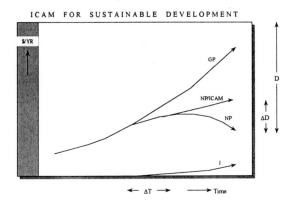

ICAM and sustainable development. The upper line (GP), which represents total gross production in the coastal area, increases as a result of population growth, technological progress and economic development. However, the damage (D) also increases due to accelerated sea-level rise, pollution, congestion, over exploitation of resources, and others. If no ICAM programme is adopted, the increase in damage will be higher than the increase in total gross production, resulting in reduced net production (NP). Reduction of damage (D) can be achieved by proactive investments in ICAM (I). Due to the long lead times required for the measures to have an effect, it will take some time before an actual reduction in damage is realised. This time lag is indicated by T. During this period, no benefits will arise from the investments. Later, however, the reduction in damage enables net production (NP/ICAM) to continue to increase, which can be regarded as sustainable development.

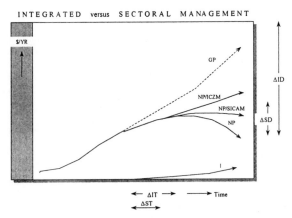

Integrated versus sectoral coastal management. The dashed line (GP) represents gross production, as in the above figure. The lower line (NP), representing net production, results from the gross production and the damage when no integrated management programme is undertaken. Reduction of damage can be achieved by proactive investments in sectoral (SD) or integrated coastal management (ID). The time lag between the start of investments is larger for integrated (IT) than for sectoral (ST) coastal management because the former involves more extensive analysis and planning. However, net production with ICAM (NP/ICAM) will eventually be higher than net production with sectoral coastal management (NP/SICAM), again because of the external effects.

Figure 10.2. The basic options for coastal communities regarding the adoption of ICAM programmes. Adapted from IPCC (1993:20–21).

TABLE 10.3.   Integrated coastal management. Triggering factors [*]

Triggering factors	Examples
	Impact issues *from human pressure and activities*
Coastal landuse and water management;	impacts brought about by tourism development exceeding the local carrying capacity;
Specific activities;	impacts brought about by filling in of wetlands to make room for seaport terminals and storage areas;
Change in environmental conditions;	morphogenetic change of the estuary's ecosystem due to navigation development;
Impact of social concern;	decreasing of fishery yields;
	Hazard issues
Shoreline erosion;	erosion caused by the location of seaport breakwaters;
Coastal river flooding;	flooding due to the cementation of the river banks;
Ocean born storms;	increasing frequency of storms due to climate change;
Migrating dunes;	changes in dune dynamics due to the pressure of human settlements and man-made structures;
	Development needs
Fisheries;	need to implement hightech-based catchment;
Protected areas and species;	need to create marine parks and to protect endangered species also with the aim of maintaining a catalyst role for tourists;
Water supply;	need to implement water supply when aquifers are altered by marine erosion;
Tourism development;	need to increase the catalysing role of beaches and other coastal sites for tourists;
Port development;	need to develop containerised seaport terminals;
Energy development;	need to implement energy supply without increasing air pollution and discharge of warm water from power plants to the salt waters;
Industrial siting;	need to reshape the existing Maritime Industrial Development Area (MIDA);
Mariculture development;	need to create a mariculture area and associated processing plants with the aim of providing complementary activities to the decreasing role of fisheries;
	Organisational process issues
Lack of co-ordination between public authorities;	inadequate co-ordination between the local branches of Ministry of the environment and the Ministry of public works;
Deficient planning and regulatory authority;	deficit of the coastal master plan;
Deficient data base and lack of information for decision-making;	unavailability of information systems providing data on the expected impacts from climate global change;
Conflicts between economic growth-aimed actions and need for environmental protection.	conflicts between local authorities and NGOs about the allocation of resource uses.

[*] Adapted from Sorensen and McCreary (1990: 38–41).

Where the social perception of the need to adopt ICAM-aimed initiatives (Fenton and Syme 1989), jointly with the interest from decision-makers in this field, diffuse and reinforce, the steering bodies are stimulated to set up and present *ad hoc* views of the coastal system. When these views focus more on unsatisfactorily impacts from conventional coastal management approaches on human health, economic prospects and coastal landscape, they succeed more in leading the local decision-making system to undertake ICAM-aimed initiatives. An example of how this approach could be designed is given in Table 10.4.

TABLE 10.4.   Possible consequences of coastal area mismanagement [*]

Coastal Issues	Consequences		
	Human health	Economic activities	Æsthetics of coastal landscape
Depletive or destructive resource use			
overpumping of water	increased health risks due to water scarcity;	salinisation caused by salt water intrusion into the aquifers;	unsightly sink holes due to subsidence;
sand mining	increased risk of injury from storms and flooding;	loss of physical assets due to erosion;	loss of unique vistas;
Loss of habitats			
land degradation, draining of wetlands	reduced nutrition of traditional collectors; increased risk of injury from storms and flooding;	decline in fisheries; loss of market and non-market wetland products;	loss of unique vistas; loss of wildlife diversity;
Pollution, environmental degradation			
water pollution	contamination of seafood; water unsafe for recreation and bathing;	decline in aquaculture, fisheries, tourism;	offensive smell;
sedimentation	contamination of seafood; water unsafe for recreation and bathing;	decline in aquaculture, fisheries, tourism;	unsightly;
solid waste disposal	increased risks of disease transmission through various vectors;	decline in aquaculture, fisheries, tourism;	offensive smell; unsightly;

[*] Adapted from Scura, Chua, Pido and Paw (1992: 59).

When the social perception and the awareness of the decision-making centres of the need to transcend conventional approaches to coastal management is disseminated, the adoption of an ICAM programme may be initiated by emphasising the advantage by undertaking the proactive strategy. The proactive action, based on abandoning the wait-and-see principle, requires long-term prospects of coastal development and may only be designed efficiently when the integration concept is adopted. In short, there are two opposite, irreconcilable approaches: on the one hand, the wait-and-see and sectoral approach, and on the other the proactive and integrated management-based approach. Sooner or later all the coastal communities and decision-making systems will have to face this option.

The Role of Social Perception. Focusing on Sea-Level Rise
One of the major impacts expected from climate change will be sea-level variations, which were discussed in Chapter 2. However, in many coastal and island regions the social and political perception of
this process it too weak. Two main reasons justify this behaviour: on the one hand, many people do not
believe that the scenarios provided by scientists are to be shared; on the other hand, the consequences
from this process will materialise in the long-term so they will not involve present generations profoundly. Where social perception needs triggering, it could be useful to use simulation techniques,
based on GIS, to show how and to what extent settlements, facilities, beaches and other resources are
expected to be jeopardised or lost. In this respect, the following conceptual presentation by UNEP is
worthy of consideration:

"The rises in temperature and in global sea level could affect coastal and marine areas, particularly in
relation to:
-   surface and groundwater flow and river regimes (water supply availability, incidence of foods and
sediment transport);
-   movement of main water masses (waves, currents, tides, erosion of the coastline, tidal range);
-   natural ecosystems due to increased temperature and exposure to climatic extremes;
-   change in the frequency and intensity of extreme events (storms, floods, winds, draughts);
-   occupation and use of coastal land due to sea-level rise.

The impacts of such changes may include:
-   an increased seawater intrusion into the coastal aquifers;
-   further difficulties in obtaining freshwater;
-   changes in fisheries and aquaculture production;
-   increased inundation under storm conditions in unprotected coastlines and low-lying areas;
-   increased shore erosion;
-   loss of natural vegetation in marginal climatic zones or areas of poor soil;
-   possible increased risk of forest fires; and
-    alteration of the biodiversity composition and structure".
(From UNEP 1995: 13).

According to Sorensen and McCreary (1990: 31–3) the triggering action leading to
the resolution on the preparation of the ICAM programme goes through two phases:
a) the *incipient awareness*, within which the recognition of crucial issues of coastal
   management is made and presented to the local community and decision makers;
b) the *growing awareness*, persuading the community and decision makers that the
   best way to face the local issues is by adopting and implementing an ICAM programme.
   The view of Sorensen and McCreary is concerned with the national scale, i.e. with
the process leading to the adoption of the national strategy plan on ICAM. Nevertheless, it could be referred usefully to the justification process on the local scale too.
   As regards how to design justification, the option between the top-down and bottom-up processes arises at several points in the ICAM process and is particularly concerned with triggering. In short, according to the social and political contexts, the
triggering action could be addressed first to the decision-making centres—e.g. the
province or municipality executive boards—and then to the local community; or the
opposite path could be followed in order to create popular inputs to the decision-
making centres. Nevertheless, it might be best to follow both paths contextually,
gradually integrating the top-down to the bottom-up processes. Anyhow, the consensus of, and support from, the local community has to be sought as soon as possible.

STAGE II—INITIATION

If the triggering action is successful, the local decision-making system—e.g. a municipality or a province—adopts the resolution calling for the presentation of the ICAM programme proposal. At this point, the initiation phase begins. The subsequent work may be carried out efficiently if technical co-operation is established by the local scientists involved and experts from the international arena serving as consultants. The presentation of the proposal is more effective when there is more consultation between the local team and organisations involved in coastal management on various scales. Co-operation with governmental and inter-governmental organisations, as well as non-governmental organisations endowed with expertise in the field may be welcome.

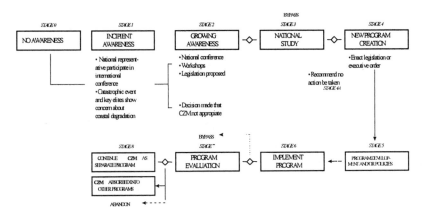

Figure 10.3. The ICAM process, according to Sorensen and McCreary (1990: 32, adapted). Acronyms: CZM, coastal zone management.

According to UNEP (1995: 24-5) the proposal of the ICAM programme should include:
a) the analysis and presentation of the ICAM prerequisites;
b) the design and presentation of the ICAM objectives;
c) a drawing of the expected boundaries of the coastal area;
d) the guidelines for integrating the decision-making centres consistently with the objectives of the ICAM programme;
e) the breakdown of the required financial and technical resources;
f) a tentative workplan and timetable;
g) the design of a monitoring and evaluation mechanism of the ICAM process.

Where the proposal is agreed by the decision-making centres concerned the triggering, and initiation work is concluded, the third stage of the ICAM process begins with the preparation of the programme.

STAGE III—PREPARATION

The basis of the ICAM programme includes:
a) the proposal;
b) the relevant resolution adopted by the local authority, or the local decision-making system concerned; and
c) complementary materials, such as recommendations and technical orientations.

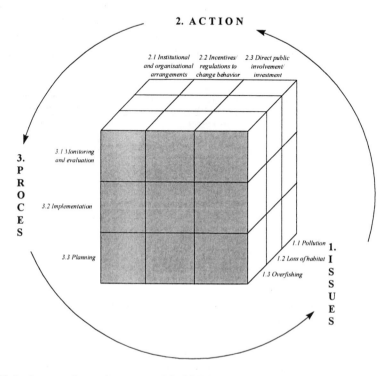

Figure 10.4. Issues, actions and processes of the ICAM process. As far as the overfishing (re issues No. 1.1) is concerned, the components of the matrix could be detailed as follows: 2. ACTIONS: 2.1 Institutional and organisational arrangements: clarification of legal rights and responsibilities; clarification of organisational jurisdiction; monitoring and enforcement; 2.2. Incentives/regulations to change behaviour: tradable fishing permits or quotas; taxes, marketing board margins, license fees; non-tradable fishing permits or quotas; regulation of gear, vessels, season; 2.3 Direct public involvement/investment: restocking, fisheries enhancement, research and development, technical assistance, education and public awareness, alternative livelihood programs, protected marine areas; 3. Processes: 3.1 Monitoring and evaluation: conduct periodically during both planning and implementation; follow progress, assess performance toward meeting objectives; identify unanticipated, social, economic and environmental consequences of management actions; recommend adjustments in actions and objectives; 3.2 Implementation: integration of essential issues and development plans; execution of activities; co-ordination of relevant actions; local participation; 3.3 Planning: proper identification and prioritisation of management issues; local participation; formulation of management strategies and policy options; formulation of action plans for implementation, monitoring and evaluation processes; evaluation, decision and adoption. Adapted from Hulfschmidt (1986) and Chua (1992).

Furthermore, the materials dating back to the triggering stage may be useful. The programme preparation obviously plays a significant role in the ICAM process. To be successful, the programme should enable and facilitate the development of the subsequent ICAM stages, i.e. planning, implementation, monitoring and evaluation. This work should generally respond to the following range of questions, which should mostly be tentatively presented and discussed in the proposal:

a) what processes and issues involving the coastal system need to be focused on;
b) what goals should be pursued through the ICAM programme;
c) which basic actions may be designed to pursue the goals;
d) how is the external environment expected to influence the ICAM process;
e) which outcomes are expected to be achieved, in the mid- and long-term, by ICAM programme implementation;
f) how should the boundaries of the coastal area be determined and traced;
g) which key investigations are needed, through which criteria are they to be carried out, and which tools are expected to be in use;
h) what expertise is needed;
i) what financial resources are needed;
j) what co-ordination mechanisms should be established;
k) what workplan and timetable may be adopted.

Processes and Issues

Being influenced by its historical roots, according to which coastal management was basically referred to environmental issues, literature has paid much more attention to physical and chemical processes involving the local ecosystem than to the socio-economic contexts. There is no doubt that, since its initial phase, i.e. in the course of the 1970s, management programmes have been conceived as useful tools not only to protect the local environment but also to pursue economic development. Nevertheless, analyses supporting the individual programmes have frequently considered the economic conditions as something less important than the crucial environmental issues, such as coastal erosion and pollution, and, more frequently, they have not adopted holistic views of the economic organisation. Consequently, the consideration of the coastal areas according to the complexity concept may be found only rarely in both literature and investigations carried out during the preparation and implementation stages of programmes. Moreover, according to Agenda 21, Chapter 17, and more substantially when the Rio spirit is considered, also the social equity principle should be regarded as inspiring the management programme, which leads to the consideration of social conditions and cultural heritage. To be fully consistent with the approach from Agenda 21 and epistemology of complexity, the coastal conditions relevant to management programmes may be imagined as an extended web of processes and issues, which may be tentatively presented by a matrix such as that, for example, shown in Table 10.5. Its content is rather more concerned with the subjects discussed in Chapters 2 (global change), 3 (ecosystem and coastal system) and 5 (coastal organisation).

TABLE 10.5.   Coastal process/issue matrix [*]

Processes	Ecosystem		Uses					
	AB	BI	AG	FA	IN	SN	TR	US
Ecosystem-concerned processes								
Sea-level rise	□	□				□		□
Subsidence	□	□			□	□		□
Eutrophication		□		□			□	
Earthquakes	□				□	□		□
Economic processes								
Containerisation						□		
Declining heavy industry	□				□	□		□
Waterfront revitalisation					□	□	□	□
Aquaculture development		□		□	□			
Social processes								
Immigration			□	□	□			□
Multi-ethnical development								□
Urbanisation	□	□	□		□		□	□
Cultural heritage protection	□	□	□		□	□	□	□

Codes:
Ecosystem: AB, *AB*iotic components; *BI*otic components;
Uses: AG, *AG*ricolture; FA, *F*isheries and *A*quaculture; IN, *IN*dustry; SN, *S*eaport and *N*avigation; TR, *T*ourism and *R*ecreational uses; US *U*rban *S*ettlements.
[*]   This matrix is not comprehensive but its role is only to show the logical exercise.

Designing Goals

The social awareness of, and social participation in the coastal management process largely depends on how the objectives of management programmes are discussed and formulated. As was discussed in Chapter 9.2, the adoption of the concept of sustainable development implies considering two goal levels: the *higher level*, where the integrity of ecosystem, economic efficiency and social equity are located; the *lower level*, where the local goals, to a varying degree connected to the higher level-referred goals, are perceived as essential for the future of the local community and, as a consequence, are regarded as pertinent to the coastal management programme. Accordingly, two aspects of the programmes may be considered. First, the *hierarchical framework* of objectives, since the individual programme goals are to be designed as an operation, on the local scale, of the final objective consisting of coastal sustainable development. Secondly, the *prioritising* of local goals, which is needed to optimise the programme implementation and, particularly, to trigger social participation. In this sense, the management programme becomes much more visible, and may be evaluated much more efficiently than in those cases where the management goals are not critically assumed, and not based on a long-run perspective. Table 10.6 shows an example of a hierarchical tree of objectives concerned with the protection of the ecosystem's integrity.

TABLE 10.6.   Example of tree of goals concerned with the protection of the coastal ecosystem

Ranks	Goals
First rank *canonical goal*	integrity of the ecosystem;
Second rank *local general goal*	protection of marine biodiversity;
Third rank *local specific goals*	a) to restore the seabed vegetation (e.g. *Posidonia Oceanica* meadows); b) to abate pollutants; c) to abate man-produced nutrients; d) to re-plan man-made physical structures facing the coastline; e). to deal with overfishing.

The Role of the External Environment

The prospect of carrying out the ICAM programme essentially depends on how the coastal system interacts with the external environment during its mid- and long-term evolution. Therefore, attention should be centred on the external environment, which is regarded as consisting of two components:

a) the *natural external environment*, consisting of the Earth's ecosystem and the large ecosystem with which the ecosystem of the individual coastal area interacts;

b) the *socio-economic external environment*, characterised by the economic and social processes on global, regional (multi-national), and national scales which are somewhat connected with the evolution of the coastal system.

In the former case the group of processes of which global change consists are the reference basis. In the latter case, globalisation of economies (global scale) and the consequences from multi-national co-operation (such as the Barcelona Convention for the Mediterranean coastal areas) are to be considered. In both cases, scenarios merit being sketched in order to draw the set of pathways which may be followed by the coastal system according to how it interacts with its external environment.

Roles and Activities of the ICAM Programme

Cicin-Sain and Knecht (1998: 46–50) suggest functions and activities of the ICAM programme being listed and classified according to how the coastal development goals are defined. Although it is concerned with the national scale, their approach is also useful in setting up programmes on the subnational scale. Where it is adapted to the local scale, major functions of ICAM programmes may be identified:

a) *area planning*: making plans for the uses of coastal resources;

b) *economic development*: encouraging development of coastal resources consistent with the sustainable development principle;

c) *stewardship rationalisation*: designing boundaries of the coastal area apt to optimise the protection of the ecosystem;

d) *proprietorship rationalisation*: establishing appropriate proprietorship regime of submerged land and waters in the responsibility of national government and local organisations;

e) *protection of public health* from man-provoked diseases and natural hazards;

f) *conflict resolution*: harmonising existing coastal uses and preventing conflicts between uses.

Each function can be complemented with the list of relevant activities. *Economic development*, for example, might be concerned with industrial fisheries, artisanal fisheries, mass tourism, ecotourism, marine aquaculture, marine transportation, port development, marine recreation, offshore minerals, ocean research, or access to genetic resources.

Actions

At this point attention shifts from the framework of goals to the actions needed to pursue them. In this respect the methodological approach by Hufshmidt (1986; also in Scura, Chua, Pido and Paw, 1992: 29) could be useful. It is based on a three-axes matrices presenting (i) process, (ii) issues, and (iii) subsequent actions. The matrix could be set up referring to the tree of goals which has just been considered. As an example, moving from Table 10.6 a matrix concerned with the protection of marine biodiversity could be set up where processes, issues and actions are considered in a general way. On the basis of this general approach, each specific goal can be considered setting up a single, dedicated matrix. This approach is visualised in Figure 10.4 where overfishing and loss of habitats (local specific goals) may be regarded as components of the protection of marine biodiversity (local general goal. As can be seen, the major usefulness of this approach consists in the possibility of correctly presenting the main characteristics of the ICAM programme.

The Boundaries of the Coastal Area

The design of the goals enables coastal managers to trace the boundaries of the coastal area. The authors quoted in Chapter 6 demonstrate, indirectly, how so much literature has paid attention to this subject. Sorensen and McCreary (1990: 8 and 28), and Scura, Chua, Pido and Paw (1992: 32) presented good views of the potential extents of the coastal area (see Figures 6.1 to 6.4). A critical consideration of these views leads us to recognise that the criteria by which coastal areas have been delimited vary remarkably from state to state according to a wide range of factors: national legislation, economic interests, crucial environmental issues, scientific influence, and others. In Chapter 6.4 it was noted that there are three main elements on the basis of which the boundaries of the coastal areas may be traced:

a) the extent of the ecosystem, or that of a group of contiguous ecosystems;
b) the administrative areas, reflecting the stewardship of the public decision-making centres;
c) the maritime jurisdictional zones.

Optimum management can be achieved where the boundaries of the coastal area coincide with those of the local ecosystem, or the set of contiguous ecosystems. Nevertheless, this is very difficult to do and the boundaries usually result from a compromise between various factors and reflect the local political context. This area challenges experts.

Key Investigations

This logical avenue leads to focus on the investigations needed to assess issues and processes, to pursue goals and undertake actions. Literature has provided many lists of subjects which should include the so-called "diagnostic of the coastal area". Three approaches, respectively by Scura, Chua, Pido and Paw (1992:35), OECD (1993a), and Cicin-Sain and Knecht (1998: 199–200) are shown in Table 10.7. They may only be regarded as reference bases because each ICAM programme needs its own list of key investigations. The lesson to be drawn from these contributions is that it is worth concentrating on how it can be gathered and processed. This issue is discussed extensively by NOAA (1992: 31–78) which, among other things, points out that:

a) data is to be tailored to the questions which investigations are required to answer;
b) qualitative data is needed where the objectives are qualitative;
c) the target year (e.g., 2025 according to the Mediterranean Action Plan) is to be determined with close reference to the objectives;
d) the geographical area to which data is to be referred may vary according to the objective.

This approach makes us think that, basically, data is to be gathered according to three ranges of variables: (i) the geographical scale, (ii) the time scale, and (iii) the methodological role (projective, prospective, goal-oriented investigations). A relevant framework is tentatively given in Table 10.8. Investigations on the ICAM programme should not just be grouped according to the mentioned variables but the tools used to collect data should be specified too which, generally, may include:

a) numerical data (statistical directories, gathering *in situ*, etc.);
b) qualitative data (from public opinion pools, legislative and regulatory materials and other kinds of investigation);
c) data from conventional maps (atlases and others);
d) remote-sensing and aerial survey data.

The data collected through various tools and presented in various formats (numerical, qualitative, maps) needs to be integrated and conveyed into holistic views. At the present state of the art, the most useful tools for this exercise are the Geographical Information Systems (GIS), whose role is expected to acquire increasing importance. Methodologies and techniques on which this approach is based are gaining increasingly sophisticated features (Clark 1996: 314–7, 405–6).

Expertise

The ICAM programme should also include the kinds of expertise needed to carry out planning, implementation, monitoring and evaluation. The team of experts vary from area to area depending on the objectives pursued in each of them. However, it should include members "experienced in (a) integrating economics, technology, ecology, and institutions, and (b) formatting and presenting information to decision-makers" (NOAA 1992: 45). The breakdown by Scura, Chua, Pido and Paw (1992: 29), based on two categories of experts (core and support experts) shown in Table 10.9, may be regarded as a possible reference basis.

TABLE 10.7. Diagnostic framework of ICAM investigations

Scura, Chua, Pido and Paw (1992: 35)	OECD (1993: 61)	Cicin-Sain and Knecht (1998:199–200)
Biophysical and environmental aspects	Processes and resources	Coastal resource bases
1. Resource inventories	1. Total area of concern	1. Existing coastal resources
2. Determination of environmental linkages and processes	2. Coastal processes occurring	2. Present use of coastal resources
3. Identification, monitoring and evaluation of environmental change	3. Influence of catchment and river systems entering the coastal area	3. Present status of coastal resources
4. Physical quantification of environmental impacts	4. Marine resources, including fauna and flora, by type and species	4. Potential for present and future use
Social and economic aspects	5. Recreation and tourism resource	Social organisation in the coastal area
1. Social, cultural and economic characterisation of coastal communities	6. Land suitable for industrial and urban development	1. Existence and character of human settlements
2. Estimation of demand and supply of coastal resources, and projection of future demand and supply	7. Drinking water supply resources	2. Economic basis for human settlements
3. Identification of current and potential future resource conflicts	The natural system: natural resource_use	3. Existence of indigenous peoples and their traditional coastal activities
4. Identification of market and policy failures	1. Area of underdeveloped land, including agricultural land	4. Social issues
5. Economic valuation of coastal resources including non market evaluation	2. Natural habitats	Existing environmental and resource-related programmes
6. Evaluation of alternative policy options and management strategies	3. Uses of forest resources	1. Environmental regulatory programmes
Institutional and organisational aspects	4. Use of water supply resources	2. Fisheries management programmes; other resource management programmes
1. Rights and obligations with regard to coastal resource use	5. Wetlands areas	3. Protected areas programmes
2. Organisational jurisdiction, responsibilities, structure and co-ordination	6. Use of coastal rivers suitable for recreation	4. Beach erosion management programmes
Opportunities For Management Interventions	7. Cultural, historical and spiritual features	5. Pollution control programmes
1. Evaluation of opportunities for, and efficacy interventions to influence behaviour	8. Use of coastal waters for shipping recreation, solid and liquid waste disposal	6. Other environmental management programmes
2. Evaluation of opportunities for, and efficacy of direct public involvement or investment	9. Water for cooling of industrial plants, fisheries, oil exploration and others	Institutional, Legal And Financial Capacity
	Impacts of development and economic activity	1. Relevant national-level institutions
	This category includes all types of waste	2. Relevant regional/provincial-level institutions
	Economic impacts of coastal area development	3. Relevant local institutions
	1. Economic benefits	4. Survey of legal authorities relative to coastal and ocean activities
	2. Costs of infrastructure investment	5. Existing capacity-building efforts
	3. Costs of development projects	
	4. Operation and maintenance costs	

TABLE 10.8. Main features of the ICAM investigations

Administrative scale	Time scale	Type of analysis	Value of results	Core methods
Meso-area (e.g. province)	Long-term (e.g. 2025)	Projective (e.g. trends on population)	Deterministic	Equation; linear functions
Mini-area (e.g. municipality)	Mid-term (e.g. 2010)	Prospective (e.g., evolution of maritime industrial development Areas, MIDAs)	Not deterministic	Scenario; stochastic equation
Micro-area (e.g., part of the municipal territory)	Short-term (e.g. 2000)	Programme-oriented (e.g., investment in coastal conservation)	Deterministic	Equations; input/output analysis and others

TABLE 10.9.   Possible core experts for ICAM

Core experts	Main responsibility
1. coastal area planner	team leadership; planning direction; general land and water uses; institutional and organisational arrangements;
2. resource economist	macroeconomic policies and their implications; economic valuation; benefit-cost analysis of management options;
3. ecologist	resource assessment; ecological impacts from coastal use development;
4. sociologist	social and cultural issues; social consultation and participation;
5. environmental engineer	technical and physical interventions; carrying capacity of coastal areas;
Support experts	Specific concern
6. parks and tourism specialist	tourism-related issues, including marine parks and nature reserves;
7. pollution expert	environmental impacts; waste management;
8. fishery expert	fishery resource assessment; harvesting and utilisation;
9. aquaculture expert	fish farming system; aquaculture planning and management;
10. forestry expert	coastal forest planning and management.

Financial and Co-ordination Mechanisms

To carry out the ICAM programme three types of "financial requirements are generally essential (...): financing of the administrative structure, the planning, the information system and the project review mechanism; financing the infrastructure and pollution control expenditures; and financing of conservation measures" (UNEP 1995: 36). Jointly with the breakdown of financial mechanisms, the programme is required to design the mechanism co-ordination needed for ICAM. According to The World Bank (1993) the main role of this mechanism is to:

a) promote and strengthen intersectoral collaboration;
b) reduce rivalry and conflicts between decision-making centres;
c) minimise duplication of functions of different decision-making centres;
d) provide a mechanism for conflict prevention and resolution between decision-makers;
e) monitor and evaluate the progress of the ICAM programme.

Workplan and Timetable

The design of the workplan, complemented by the timetable, is the final step of the ICAM programme. It should be detailed and is required to specify "which analyses are to be done by whom, in what sequence, and at what costs. The work can be depicted in various ways, such as by a bar chart or by a critical path method (CPM) diagram" (NOAA 1992: 44).

Conclusion of the Preparation Stage

During the previous stages the ICAM process has gone through two bifurcations, each of them characterised by important options for the decision-making system. The first bifurcation occurred at the end of the triggering stage when the decision-making apparatus decided to reject/accept the idea of calling for a proposal concerning the preparation of the ICAM programme. The second bifurcation took place at the end of the justification stage and consisted in adopting/rejecting the ICAM programme proposal. The third bifurcation occurs now, at the conclusion of the preparation stage, when the decision-making system had to take the resolution relevant to the adoption/rejection of the ICAM programme.

STAGE IV—PLANNING

If the programme is adopted, the fourth stage of the ICAM process starts up by focusing on planning. *Sensu lato*, it can be considered as consisting of three closely integrated components: (i) designing strategies, (ii) zoning, and (iii) planning.

Designing Strategies

Strategy basically consists of a set of guidelines illustrating the goals of the ICAM process and how actions included in the master, sectoral and special plans should be carried out. The framework of these materials varies according to the coastal areas as a result of the features of their social and ecological contexts, the processes by which they are influenced and the interaction with their external environment. Table 10.10 shows two different approaches as a basis for discussion by OECD/UNEP and Sorensen and McCreary, respectively.

TABLE 10.10.   ICAM strategies

OECD and UNEP	SORENSEN and McCREARY
Strategies pertaining to individual economic sectors	1. inventory of natural resources
1. management of fishery resources	2. human settlements
2. management of recreation and tourism	3. land use and development allocation
3. management of extraction of minerals, sand, gravel, oil and gas	4. environmental consideration in project planning
4. management of transport	5. food production and raw materials
	6. conservation and environmental protection
Strategies pertaining cross sectoral activities	7. recreation and tourism
1. pollution control	8. infrastructures and engineering work
2. management of coastal landuse	9. construction materials
3. protection of significant landscapes	10. public health
4. conservation of coastal ecosystems	11. management of water resources
	12. institutional framework
	13. navigation, shipping and harbours
	14. security

*   Adapted from OECD (1993a: 70) and UNEP (1995: 30).
**   Adapted from Sorensen and McCreary (1990: 79).

Planning

Planning is the logical consequence of designing strategies and essentially includes:
a) the *master plan*, dealing with the basic features of the programme;

b) *sectoral plans*, dealing with the individual sectors of the coastal system (seaports, navigation, agriculture, industrial fisheries, artisanal fisheries, aquaculture, etc.);
c) *special plans*, dealing with cross-sectoral issues and objectives of special relevance to the pursuit of coastal sustainable development (plans for protected areas, contingency plans, etc.).

Obviously, each coastal area, according to the features and issues by which is marked, needs its own endowment of sectoral and special plans.

The list of sectoral plans as well as, to some extent, the list of special plans may usefully be tailored to the coastal use framework which was presented in Chapter 7.4 (Table 7.4). If so, sectoral plans are concerned with selected categories of coastal uses included in the coastal use framework in such a way as to optimise consistency between the method through which the coastal organisation is assessed and the framework of the plans on which the coastal programme is based. Other approaches are clearly possible, such as that by Sorensen and McCreary shown in Table 10.11.

TABLE 10.11.   Sector planning in the ICAM programme

Specific sectors of the coastal area	Rarely specific sectors of the coastal area but having direct impact
Navy and defence operations	agriculture
Port and harbour development	forestry
Shipping and navigation	fish and wildlife management
Recreational boating and harbours	education
Commercial and recreational fishing	housing
Mariculture	water supply
Tourism	transportation
Marine and coastal research	flood control
Shoreline erosion control	oil and gas development
	mining
	industrial development
	energy generation
Source: Sorensen and McCreary (1990: 89).	

Actions included in plans are likely to be more effective when they are illustrated in detail. Each action could be characterised by a "task presentation" embracing: (i) the description of the target or product, (ii) the description of methodology, (iii) the duration, (iv) the team leader, and (v) the institutions involved.

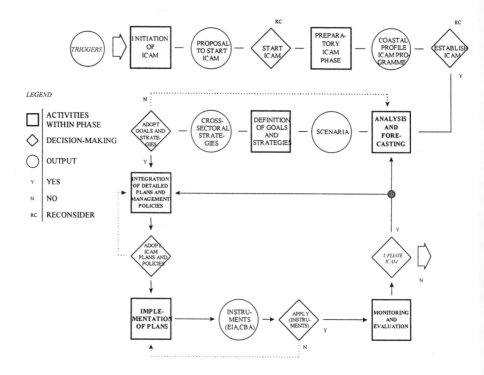

Figure 10.5. The ICAM process, according to UNEP (1995: 22, adapted).

Zoning

The coastal area is geographically subdivided through zoning. Each zone is included in the ICAM programme by indicating the coastal uses which may be developed, those which are prohibited, and by specifying restrictions, incentives, guidelines, and other institutional and functional elements designing its role in the context of the coastal area. This operation, if done properly, can give a strong impetus to the correct, sustainable use of coastal resources. The used spectrum of methodologies and techniques from urban and regional planning provide a useful basis for dealing with this task. Zoning can provide two other basic advantages. First, an operational advantage because the organisation of the coastal area is reflected in a well-defined spatial setting, thereby clarifying the expected outcomes from the ICAM process on the individual, marine and land sites and spaces and enabling the local community to discuss it correctly. Secondly, a methodological advantage because, in this context, coastal planning acquires similar features to those of regional planning and, as a result, may be approached by concepts and methods provided by regional theory.

STAGE V—IMPLEMENTATION

Stage IV leads to another bifurcation, i.e. the option between the adopting or rejecting plans, zoning and strategies. If this stage is marked by a positive reaction from the decision-making system and the local community, the ICAM programme may start to be implemented. In principle, this is a long term process, characterised by a set of targets which should each be achieved through the individual actions prefigured by plans and guidelines.

TABLE 10.12.   The ICAM process. Components of planning and implementation

Planning	Implementation
1. identifying issues and problems, and establishing corresponding objectives and criteria;	1. ensuring concordance between the plan structure and its implementation;
2. delimiting the spatial, temporal and substantive scopes of the planning effort;	2. designing, constructing, operating and maintaining physical structures;
3. identifying social actors, and ensuring their participation in the management process;	3. applying and modifying regulatory measures, such as physical planning;
4. analysing existing planning programmes, institutional arrangements and management instruments to determine whether they may be useful for addressing the issues;	4. applying and modifying standards relating to, for example, water quality;
5. formulating a set of actions that systematically relate the ICAM objectives to the current state of the coastal system;	5. enforcing strategies, regulations and standards through a formal legal process, or through persuasion, education and traditional community controls;
6. collecting and analysing existing data and evaluating the need for further research and information;	8. providing for the participation of private entities and the public;
7. establishing monitoring systems and integrated databases;	7. identifying and contracting sources of funding for the implementation process;
8. supplying information for programme evaluation to policy makers;	
Adapted from World Coast Conference (1993: 27).	

Literature has paid ample attention to all the aspects and issues of the implementation stage because the programme is more effective when there are more efficient tools and mechanisms. Although there is no room to synthesise this wide spectrum of contributions, the following framework of implementation tools, on which many authors have agreed, can be regarded as a good starting point in designing the ICAM implementation mechanisms.

Regulatory Instruments

Commonly referred to as a "command and control philosophy", regulatory instruments can be characterised as "institutional measures (on the basis of come form of legislation) aimed at directly influencing the environmental management" (OECD 1989: 12). "The range of issues covered by regulatory instruments includes land-use planning, building regulations, construction guidelines for the coastline, conservation regulations, standards for anti-fouling for boats, fishing quotas, marine transport regulations, aquaculture requirements and licensing of various other activities" (UNEP 1995: 47).

Economic Instruments

Literature is not fully in agreement on how economic instruments may be defined and grouped. According to the approach by OECD (1993a: 67–8), on which UNEP's approach (1995: 48–9) is also based, economic instruments may be placed in five groups:

a) *charges* (effluent charges, user charges, product charges, administrative charges, tax differentiation);
b) *subsidies* (grants, soft loans, tax allowances);
c) *deposit-refund systems*;
d) *market creation* (such as emissions trading, market intervention, liability insurance);
e) *financial enforcement incentives* (non-compliance fees, performance bonds).

Bargaining and Negotiation

These tools may positively influence the ICAM process. They can be used with industry "in conjunction with either regulatory or economic instruments. This method is becoming ever more frequent in dealing with one particular industry group or with an industrial association on a regional basis. Specific agreements with industry or other groups, e.g. tourism associations, can be particularly useful instruments in coastal areas where more stringent pollution or conservation standards might be needed" (UNEP 1995: 49).

Conflict Resolution

This topic was widely discussed in Chapter 8 where the procedures of preventing and resolving conflicts were presented. It is sufficient here to recall that two basic groups of instruments are applied: arbitration and legal procedures. The former, which can be conducted through various criteria, is the preferable tool. The legal instruments should be only applied where all the arbitration procedures have failed. The ICAM programmes, by their nature, are required to give all their attention to this subject. When conflicts can be resolved without legal procedure the ICAM programme may be implemented more effectively.

Education and Training

The effectiveness of the ICAM programme implementation largely depends on education and training. Special plans are included in the programme and wide co-operation is stimulated by academic and extra-academic bodies, including non governmental organisations endowed with scientific and technical resources. In this respect education and training may be regarded as the major components of the human resources development and capacity building (Scura Chua Pido and Paw 1992: 82–6).

Social Communication
This expression does not occur in literature. It is presented here with the aim of embracing all the information networks, mass communication tools and any other tools which can be used to disseminate information on an ICAM process within the local community, to strengthen and widen the social perception of the need for, and advantages from, this process, and to catalyse social participation and collaboration.

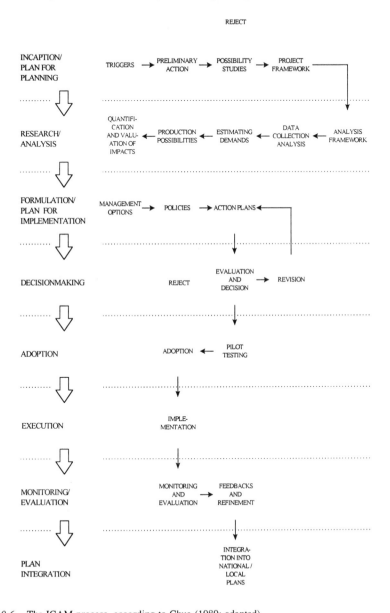

Figure 10.6. The ICAM process, according to Chua (1989: adapted).

STAGE VI—MONITORING AND EVALUATION

The ICAM programme needs to be monitored during implementation. This work provides useful information for evaluating the effectiveness of the ICAM process. Accordingly, the ICAM programme is expected to include *ad hoc* phases during which the local decision-making system evaluates the outcomes of the programme implementation. On the basis of evaluation, the programme is confirmed or abandoned. If confirmed, it may be revised in order to tailor it to changing needs and influences from the external environment.

Monitoring may be concerned with two distinct, but complementary, tasks. On the one hand, it assesses the environmental, economic and social impacts from coastal management in order to set up a view of how the coastal system evolves. On the other hand, it describes how the ICAM process progresses and adjusts scenarios step by step.

Monitoring Coastal Impacts
This subject should be dealt with starting from the final objective of coastal management, namely sustainable coastal development. In this respect, it is methodologically correct to refer the techniques of monitoring coastal impacts to the three components of this objective: (i) integrity of the ecosystem, (ii) economic efficiency, and (iii) social equity. A possible breakdown of techniques is presented in Table 10.13. It would be inappropriate to discuss these techniques in detail here, so only a few considerations on their roles are mentioned.

TABLE 10.13.   Techniques for monitoring coastal impacts

Techniques	Components		
	Quantitative data	Qualitative data	GIS
Environmental impacts			
Rapid Assessment of Coastal Environment, RACE	❏	❏	
Environmental Impact Assessment, EIA	❏	❏	
Impacts from Expected Climate Change, IECC	❏		❏
Natural Risk Assessment, NRA	❏		❏
Economic impacts			
Cost-Benefit Analysis, CBA	❏		
Economic Impact Assessment, EIA	❏		
Carrying Capacity Assessment, CCA	❏	❏	❏
Economic Growth Analysis, EGA	❏		
Social impacts			
Social Impact Assessment, SIA		❏	❏
Human Development Indicator, HDI		❏	❏

Environmental Impacts
They may be monitored by four procedures.

The *Rapid Assessment of Coastal Environment* (RACE) is a method proposed by Scura, Chua, Pido and Paw (1992: 36) consisting of a "fast procedure for analysing the socioeconomic and environmental conditions of a given coastal area. Rapid appraisal techniques utilise many research tools similar to those employed in conventional research. However, to speed the analysis, the focus of the appraisal is on a narrower set of characteristics or key indicators". Its detailed description can be found in Chua and Pido (1992: 137–47).

The *Environmental Impact Assessment* (EIA) is a "tool used in many coastal management contexts. In fact, it is one of the mainstays of many nations" environmental protection program. It is a process whereby developers, or others supporting or promoting a project, conduct an analysis to determine the likely impacts of the proposed project or activity. The process originated in the Unites States and was institutionalized in the National Environmental Policy Act of 1969. An important part of the EIA process is the determination of alternative approaches to the project which will reduce (or eliminate) adverse environmental impacts" (Cicin-Sain and Knecht 1996: 186). This tool can be used either for micro-projects, such as that of thermoelectric plants, or a wide project, such as that of a manufacturing area. "The complex procedures used in some developed countries (as an example, in the member states of the EU) could be difficult to apply in developing countries (such as those of Mashrek) due to the extensive data requirements and institutional and technical requirements needed for their implementation, the considerable time involved and the consequent costs in their operation" (UNEP 1995: 41). Useful discussion can be found in Sorensen and McCreary (1990: 72–7) and a presentation of the procedure in Clark (1954: 118–30).

Impacts from Expected Climate Change is a procedure recommended by UNEP (1995: 43) to analyse the possible impacts that climate change, and subsequent sea-level rise and acceleration of coastal erosion, are expected to bring forth in coastal areas. The final goal of analysis is to design how man-made structures should be adapted and what other measures would be useful to deal with these physical processes.

The *Natural Risk Assessment* (NRA) is a well known procedure aimed at evaluating the human-induced and natural risks which coastal populations could undergo. Risks vary according to the latitudes and the tectonic features of the region. As an example, in the Mediterranean the major risks are earthquakes, volcanism, river flooding and disasters provoked by river and marine erosion.

Economic Impacts
They may be monitored by three procedures.

Cost-Benefit Analysis (CBA) is a well-known procedure, illustrated in many economic textbooks. Basically, it was conceived to evaluate the benefits/damage concerned with the project of an individual man-made structure, or a wide project, such as the revitalisation of a maritime waterfront. Evaluation is made in financial and economic terms and should take into account the environmental risks that the structures investigated are susceptible to cause. In this respect the procedure is linked to the Natural Risk Assessment (UNEP 1995: 45-6).

Carrying Capacity Assessment (CCA). This procedure has spread out in recent times. It "can be defined as the maximum load of activity or maximum number of users which can be sustained by a natural or man-made resource of system without endangering the character of that resource. By defining the carrying capacity for a certain activity it is possible to establish the framework for development and management in the areas under consideration" (UNEP 1995: 43).

Economic Growth Analysis (EGA) is proposed with the aim of providing a global view of how speedily, and along which avenues, the local economic system has developed. Analysis may be based on the use of *Gross Area Product* (GAP), or the *Gross Product per Capita* (GPC), as well as other indicators pertaining to the wealth which has been produced in the coastal area from the starting up of the ICAM programme. Also comparisons between the area under evaluation and other coastal areas, benefiting or not from ICAM programmes, may be useful.

Social Impacts

Literature on ICAM has not focused in depth on the social impacts. However, bearing in mind that integrated management is the means through which sustainable development is pursued, it also follows that the social dimension needs to be monitored. Here an approach and a complementary indicator is proposed.

Social Impact Assessment. Moving from the concept of human development, as it was defined by the United Nations Development Programmes (UNDP), the social impacts brought about by the ICAM process are analysed with the aim of evaluating whether and how social equity is implemented while ICAM process is evolving. The approach developed by UNDP through the *Human Development Reports* may be the basis moving from which the state of education, water supply, health, safeguarding of human and political rights, access to the natural and cultural heritage, safeguarding of children, women and minorities, among others, are monitored and framed in a global view. As regards the coastal area this approach is also useful to design possible social impacts brought about by the projects and programmes, including sectoral ones which are under evaluation.

Human Development Indicator (HDI). As is well known, this is a synthetic indicator used by UNDP to provide a measure of the extent to which human development is pursued. While the Gross Product per Capita is an economic indicator, HDI is a social indicator. By calculating the HDI on the local scale it is possible to complement the analysis of human development of the coastal area with a measure providing a global perception of the social trends of coastal management.

Monitoring the Coastal Management Process
The above mentioned procedures and indicators are aimed at monitoring possible evolutions of the coastal system as a result of the development of projects and programmes. Nevertheless, it is also necessary to monitor whether the ICAM programme is proceeding according to the workplan and timetable adopted in stage III (Preparation) of the ICAM's process. This subject is emphasised by Sorensen and McCreary (1990: 113–26), who presented this set of indicators of the ICAM progress:
a) budget allocation per year;
b) number of permits issued, denied, conditional;
c) consistency of law dealing with coastal management;
d) number of agreements or memoranda executed for interagency co-operation;
e) availability of appropriately trained and educated staff;
f) number of subnational programs initiated or approved;
g) quality of information used in program development.

Evaluation of Outcomes
The effectiveness of the ICAM programme needs to be evaluated, and possibly measured by *ad hoc* indicators. Sorensen and McCreary (1990: 114) consider it useful to use two sets of indicators, respectively dealing with the instrumental factors of the ICAM programme and the evaluation of environmental and socio-economic conditions. Their approach is presented in Table 10.14.

TABLE 10.14.   Outcome indicators of the ICAM programme

Instrumental factors	Environmental and socio-economic conditions
Cost and length of time for permit review	Water quality (dissolved oxygen, nutrients levels)
Number of procedures and steps eliminated	Fishery yields
Public participation (number of individuals and groups)	Protein component of diet derived from coastal fisheries
Geographic scope and issue coverage of information base	Number and linear distance of access ways
	Kilometres of coast in public ownership
	Number of recreation user days
	Number of coastal species included in endangered species list
	Acreage of wetlands protected or restricted
	Number of low and middle income housing units provided within the coastal zone
	Tonnage and value of commodities handled in ports
	Employment derived from fisheries, ports and tourism sectors
	Hazard impacts (lives lost, property damaged)

10.3. A Political Concern

From the presentation sketched in this Chapter the ICAM process appears to be a very complicated, binding enterprise. It consists of five stages, followed by monitoring and evaluation, and each phase is technically difficult to design and carry out. Besides, in order to be successful, ICAM requires deep involvement of decision-makers at every level, and extensive social participation. Any complication is mainly due to the *conventional* regional planning being only a part of the ICAM process, practically stages III and IV. Triggering and initiation precede these stages, followed by implementation, monitoring and evaluation. As a result, the spectrum of actions and activities pertaining to the ICAM process is much wider than conventional regional planning. However, this does not imply that the ICAM approach should not have been undertaken. If innovative, integration-inspired management avenues are not crossed, the global change—bringing about sea-level rise and accelerating coastal erosion—and the associated increase of human pressure, together with the changing economic and social contexts, could drive many coastal regions into irreversible degradation jeopardising the present and, more deeply, future generations. Hence, ICAM at present is an unavoidable avenue.

This framework leads us to believe that, where ICAM programmes are adopted, close co-operation should be set up between the local decision-making systems and research bodies, on the one hand, and the national and international organisations and institutions, on the other. ICAM has become widespread throughout the world thanks to strong impulses from the UN system. Other inter-governmental organisations, including EU, have begun to undertake similar actions. This is the first time that regional planning and management on the local scale have been supported by international programmes. Thus, the ICAM programme is worth adopting because the local decision-making system may benefit from scientific and technical support from the international milieu. As a final result, the effectiveness of the ICAM process is correlated with two co-operation co-ordinates, i.e. (i) co-operation between science and policy on the local scale, and (ii) co-operation between the local decision-making systems, governmental organisations, and inter-governmental organisations.

CONCLUSION

This book seeks to demonstrate that the pursuit of sustainable coastal development through integrated management, and that of making programmes and actions sensitive to global change, induces the scientific community to reject a positivism-rooted epistemological background, disjunctive logic and analytical methods, and to adopt a complexity-based epistemology, conjunctive logic and inductive methods. The decision-making system, in its turn, is required to tailor its goals and strategies to a holistic view of present and future coastal systems. It follows that ICM is an unprecedented task.

A view of the diffusion of coastal management pioneer programmes on the national scale, almost all rather inconsistent with the integration principle, was provided by Sorensen (1993), and more recently by Cicin-Sain and Knecht (1998). Sorensen divided the ocean world into seven areas and calculated the ratio between (a) States which had undertaken ICAM programmes, and (b) the totality of existing States in each area (a:b•100). That investigation showed that only all the North American states are endowed with ICAM programmes, while the ratio is little over 50 per cent in Asia and Latin America, and it decreases to low and very low values in other parts of the world.

The breakdown by Sorensen may usefully be associated with the framework of the existing UNEP Action Plans. This operation leads us to evaluate whether and to what extent there is multi-national co-operation with coastal management programmes in regions where coastal areas are liable to increased human pressure, such as those embraced by the UNEP Regional Seas Programme. This is the resulting framework.

ICAM programmes and action plans

Geographical areas	Percentage of coastal States with ICAM programmes	UNEP Action Plans	Association between ICAM programmes and Action Plans
North America	100	lacking	not relevant
Caribbean and West Atlantic islands	31	existing (Caribbean)	very low
Central America	57	existing (Caribbean)	medium
South America	45	existing (South Pacific)	low
Europe and North Atlantic	32	existing (only the Mediterranean and Black Sea)	low
Near East	46	existing (Mediterranean and Near East)	medium
Africa and the Western Indian Ocean	13	existing (West and Central Africa)	low
Asia	57	expected (South Asia, Eastern Asia, North Western Pacific)	high in South-East and East Asia (expected)
Oceania	33	existing (South Pacific)	low

This breakdown is only slightly significant, for at least three reasons: (i) the UNEP operated Action Plans merely cover some of the geographical areas to which coastal management programmes are referred; (ii) the state of progress of the adopted Action Plans differs from area to area; (iii) only the most advanced Action Plans, such as the Mediterranean Action Plan (MAP), were also designed to pursue ICM. However, it suggests dividing the ocean world into three areas.

a) *Mature coastal areas*: all the North American coastal areas have achieved, or are about to reach, such high organisational level as to be regarded very close to operating substantial integrated management and pursuing sustainable development.

b) *Maturing coastal areas*: Central America, Eastern and South-Eastern Asia and Near East, together with some parts of the EU space, are characterised by coastal area management programmes which are acquiring the shape of ICM programmes and are diffusing.

c) *Less organised coastal areas*: except for some States, the rest of the world is far from building up effective integrated coastal management.

This cluster shows how just a restricted part of the world is actually involved in carrying out integrated coastal management. This evaluation becomes more significant if it is taken into account that, despite claiming to be tailored to the integrated management principle, many existing programmes are not sufficiently consistent with the sustainable development principle. It would therefore be justifiable to state that the task to be tackled by science and politics is more difficult than what, *prima facie*, can be deduced from data by Sorensen.

These considerations highlight a paradoxical situation. On the one hand, the need in many areas to implement and disseminate ICM is, often rhetorically, emphasised as both a necessary and easy task to meet. On the other hand, the scientific and political approaches, strategies and actions that ought to be conducted to meet this need are so inflexible that they insinuate doubts as to whether they are likely to be successful or not.

When considering the issue more closely it becomes evident that the major concern is not technical, i.e. it does not relate to the design of technical aspects such as zoning, or technical tools, such as those of monitoring and evaluation. Instead, the major concern is epistemological because the technical operations must be sustained by a profound change in the way of exploring and managing the spatial system, such as the coastal area. This reasoning leads to emphasising the pre-eminence of education. The greatest concern involves implementing and diffusing curricula on ICM aimed at providing constructivism-inspired fundamentals and associating them with the presentation of the individual key issues and the learning of relevant technical tools. The more international co-operation on global and regional (multi-national) scales conveys cultural and scientific resources into this field, the more successful the efforts are. This enterprise is not only particularly prominent because of the special nature of coastal management but also because it could provide a significant example of how, in joining their efforts and resources, science and politics might deal with the complexity of post-modern society. In this respect, ICM is rich in metaphorical values.

ACRONYMS AND ABBREVIATIONS

AAAS	American Association of Advancement of Science
BP	Blue Plan
CAMP	Coastal Area Management Programme
cb	coastal baselines
CBA	Cost-Benefit Analysis
CCA	Carrying Capacity Analysis
cm	centimetre
CM	Coastal Management
COM	Coastal Organisation Model
CPM	Critical Path Method
CS	Continental Shelf
CSD	Commission on Sustainable Development
CZ	Contiguous Zone
CZM	Coastal Zone Management
CZMA	Coastal Zone Management Act
DSS	Decision Supporting System
EC	European Commission
ECA	Economic Impact Assessment
EEC	Economic European Community
EEZ	Exclusive Economic Zone
EFZ	Exclusive Fishery Zone
EGA	Economic Growth Analysis
EIA	Environmental Impact Assessment
ELOISE	European Land-Ocean Interaction Studies
EOCMA	Economic Organisation Model of the Coastal Area
ERS	Environment Remote Sensing
ESP	Exchangeable Sodium Percentage
EU	European Union
EXP	Exploitability
FAO	Food and Agriculture Organization
ft	Feet
GAP	Gross Area Product
GCM	Global Circulation Model
GESAMP	Joint Group of Experts on the Scientific Aspects of Marine Pollution
GIS	Geographic Information Systems
GMIM	Global Marine Interaction Model
GPC	Gross Product per Capita
grt	gross registered ton
HC	[Protection of Coastal] Historic Settlements
HDEGCP	Human Dimensions of Environmental Global Change Programme
HDI	Human Development Indicator
HDP	Human Dimensions [of Environmental Global Change] Programme
IA	Industrial Area
ICAM	Integrated Coastal Area Management
ICM	Integrated Coastal Management
ICSU	International Council of Scientific Unions, now International Council for Science
IECC	Impacts from Expected Climate Change

235

IGBP	International Geosphere-Biosphere Programme
IMO	International Maritime Organization
IOC	Intergovernmental Oceanographic Commission
IPCC	Intergovernmental Panel of Climate Change
km	kilometre
LME	Large Marine Ecosystem
LOICZ	Land-Ocean Interactions in the Coastal Zone
LOS	Low of the Sea
m	meter
MAB	Man and Biosphere Programme
MAP	Mediterranean Action Plan
MARPOL	International Convention for Prevention of Pollution from Ships
MCDS	Mediterranean Commission on Sustainable Development
MCE	Mediterranean Coastal Ecosystem
MEDPOL	Mediterranean Pollution [Monitoring and Research Programme]
MFT	Mediterranean Trust Fund
mht	mean high tide
MIDA	Maritime Industrial Development Area
MLME	Mediterranean Large Marine Ecosystem
mlt	mean low tide
mm	millimetre
mm/yr	millimetre per year
NGO	Non Governmental Organisation
nm	nautical mile
NOAA	National Oceanic and Atmospheric Administration
NRA	Nature Risk Assessment
OCM	Ocean Circulation Model
OECD	Organisation for Economic Co-operation and Development
PAP	Priority Actions Programme
RAC	Regional Activity Centre
RACE	Rapid Assessment of Coastal Environment
REMPEC	Regional Marine Pollution Emergency Response Centre
ROCC	Regional Oil Combating Centre
SIA	Social Impact Assessment
SPA	Specially Protected Areas
sq km	square kilometre
TS	Territorial Sea
UN	United Nations
UNCED	United Nations Conference on Environment and Development
UNCLOS	United Nations Conference on the Law of the Sea
UNDP	United Nations Development Programme
UNEP	United Nations Environment Programme
UNESCO	United Nations Educational Scientific and Cultural Organization
yd	yard
yds	yards
yr	year

REFERENCES

General

Agardy M.T., Accomodating Ecotourism in Multiple Use Planning of Coastal and Marine Protected Areas. *Ocean and Coastal Management*, **20**(3), 1993, 219-240.

Alexander L.M., Geographic Perspectives in the Management of Large Marine Ecosystems. In Large Marine Ecosystems: Patterns, Processes and Yields, ed. K. Sherman, L.M. Alexander & B.D. Gold, American Association for Advancement Science, Washington D.C., 1990, pp. 220-223.

Alexander L.M., Large Marine Ecosystems: a new focus for marine resources management. *Marine Policy*, **17**(3), 1993, 186-198.

Alexander L.M., The Management of Enclosed and Semi-Enclosed Seas. In *Ocean Management in Global Change*, ed. P. Fabbri, Elsevier Applied Science, London, 1992, pp. 539-549.

Alexander L.M., The Role of Choke Points in the Ocean Context. *GeoJournal*, **24**(4), 1992, 503-510.

Anderson L.G., *The Economics of Fisheries Management*. J.Hopkins University Press, Baltimore, Maryland, USA, 1986.

Anderson S.H., The Role of Recreation in the Marine Environment. In *Ocean Yearbook 2*, ed E. Mann Borgese & N. Ginsburg, University of Chicago Press, Chicago, Illinois, USA, 1980, p. 183-198.

Archer J.H., Resolving Intergovernmental Conflicts in Marine Resource Management: The US Experience. *Ocean and Shoreline Management"*, **12**(3), 1989, 253-270.

Arnason R., Ocean fisheries management: recent international developments. *Marine Policy*, **17**(5), 1993, 334-339.

Bakun A. Definition of environmental variability affecting biological processes in large marine ecosystems. In *Variability and Management of Large Marine Ecosystems*, ed. K. Sherman & L.M. Alexander, American Association for Advancement Science Selected Symposium 99, Westview Press, Boulder, Colorado, USA, 1986, pp. 89-108.

Ballinger R.C., Smith H.D. & Warren L.M., The Management of the Coastal Zone of Europe. *Ocean and Coastal Management*, **22**(1), 1994, 45-85.

Barcena A., UNCED and Ocean and Coastal Management. *Ocean and Coastal Management*, **18**(1), 1992, 15-54.

Barg U.C., *Guidelines for the promotion of environmental management of coastal aquaculture development.* FAO, Rome, 1992.

Barston R.P., International Dimensions of Coastal Zone Management. *Ocean and Coastal Management*, **23**(1), 1994, 93-116.

Bascom W., Deep-water archaeology. In *Oceanography. Contemporary Readings in Ocean Sciences*, Oxford University Press, New York, 1977, pp. 269-286.

Bass G.F., Marine Archaeology: A Misunderstood Science. In *Ocean Yearbook 2*, ed E. Mann Borgese & N. Ginsburg, University Chicago Press, Chicago, Illinois, USA, 1989, 137-152.

Bax N.J.& Laevastu T., Biomass Potential of Large Marine Ecosystems: A Systems Approach. In *Large Marine Ecosystems: Patterns, Processes and Yields*, ed. K. Sherman K. L.M. Alexander & B.D. Gold, American Association for Advancement Science, Washington D.C., 1990, pp. 188-205.

Beatley T., Protecting Biodiversity in Coastal Environments: Introduction and Overview. *Coastal Management*, **19**(1), 1991, 1-19.

Becet J.-M., *L'aménagement du littoral*. Presses Universitaires de France, Paris, 1987.

Belsky M.H., Developing an ecosystem management regime for large marine ecosystems. In *Biomass Yields and Geography of Large Marine Ecosystems*, ed. K. Sherman K. & L.M. Alexander, American Association for Advancement of Science, Selected Symposium 111, Westview Press, Boulder, Colorado, USA, 1980, pp. 443-459.

Berryman K., Tectonic Processes and their Impact on the Recording of Relative Sea-level Changes. In *Sea surface studies. A global view*, ed. R.J.N. Devoy, Croom Hekm, London, 1987, pp. 127-164.

Bigford T.E., Sea-Level Rise, Nearshore Fisheries, and the Fishing Industry. *Coastal Management*, **19**(4), 1991, 417-437.

Bird E.C., *Coastline Changes. A Global Review*. Wiley & Sons, New York, 1985.

Birnie P., Delimitation of maritime boundaries: emergent legal principles and problems. In *Maritime Boundaries and Ocean Resources*, ed. G. Blake, Croom Hekm, London, 1987, pp. 15-37.

Birnie P.W., The Law of the Sea and the United Nations Conference on Environment and Development. In *Ocean Yearbook 10*, ed E. Mann Borgese, N. Ginsburg & J.R. Morgan, University Chicago Press, Chicago, Illinois, USA, 1993, p. 13-39.

Boaden P.J.S. & Seed R., *An Introduction to Coastal Ecology*. Blackie, Glasgow, UK, 1985.

Boelaert-Suominen S. & Cullinan C., *Legal and institutional aspects of integrated coastal area management in national legislation*. FAO, Rome, 1994.

Bowen R.E., The Role of Emerging Coastal Management. Practices in Port and Harbour Management. In *Ports as Nodal Points in a Global Transport system*, Proceedings of Pacem in Maribus XVIII, August 1990, ed A.J. Dolman & J. van Ettinger, Pergamon Press, Oxford, UK, 1992, pp. 229-238.

Brillat T.H. & Liffmann M., The Implications of Marpol Annex V on the Management of Ports and Coastal Communities. *Coastal Management*, **19**(3), 1991, 371-390.

Brown E.D. & Churchill R.R. ed, *The UN Convention on the Law of the Sea: Impact and Implementation*. University of Hawaii, Honolulu, Hawaii, USA, 1987.

Brown E.D., Protection of the underwater cultural heritage. Draft principles and guidelines for implementation of Article 303 of the United Nations Convention on the Law of the Sea, 1982. *Marine Policy*, **20**(4), 1996, 325-336.

Brown E.D., *Sea-bed energy and mineral resources and the law of the sea*. vol. I: *The areas within national jurisdiction*. Graham & Trotman, London, 1984.

Bruce M., Ocean Energy: Some Perspectives on Economic Viability. In *Ocean Yearbook 5*, ed E.Mann Borgese & N. Ginsburg, University Chicago Press, Chicago, Illinois, USA, 1985, p. 58-78.

Burbridge P.R., Dankers N. & Clark J.R., Multiple-Use Assessment for Coastal Management. In *Coastal Zone '89*, ed. O.T. Magoon *et al.*, American Society of Civil Engineers, New York, Vol. I, 1989, pp. 16-32.

Burke W.T., UNCED and the oceans. *Marine Policy*, **17**(6), 1993, 519-533.

Burroughs R.H., Ocean dumping. *Marine Policy*, **12**(2), 1988, 96-104.

Busha T.S., The IMO Conventions, (E.Mann Borgese & N. Ginsburg ed), University Chicago Press, Chicago, Illinois, USA, 1986, p. 160-170.

Butler M.J.A., Le Blanc C., Belbin J.A. & MacNeil J.L., *Marine resource mapping: an introductory manual*. FAO Fisheries Technical Papers, FAO, Rome, 1987.

Caddy J.F. & Bakun A., A Tentative Classification of Coastal Marine Ecosystems Based on Dominant Processes of Nutrient Supply. *Ocean and Coastal Management*, **23**(3), 1994, 201-211.

Caddy J.F. & Sharp G.D., *Un marco ecològico para la investigaciòn pesquera*. FAO Documento Tecnico de Pesca, No. 283, FAO, Rome, 1988.

Carey J.J., Mieremet R.B., Reducing Vulnerability to Sea Level Rise: International Initiatives. *Ocean and Coastal Management*, **18**(1), 1992, 161-178.

Carter R.W.G., Man's Response to Sea-level Change. In *Sea surface studies*, ed. R.J.N. Devoy, Croom Helm, London, 1987, pp. 464-498.

Cavallo A., Hock S., & Smith D., Wind energy: technology and economics. In *Renewables for fuels and electricity*, ed. T.B. Johansson *et al.*, United Nations, New York, 1992, 45-81.

Chapman D.M., Shore Protection. Conflicting Objectives in Decision Making. In *Coastal Zone '91*, ed. O.T. Magoon, H. Converse, V. Tippie, L. Th. Tobin & D. Clark, vol. III, American Society of Engineers, New York, 1991, pp. 2340-2353.

Charles A.T., Fishery conflicts: a unified framework. *Marine Policy*, **16**(5), 1992, 379-393.

Chaussade J. & Corlay J.-P., *Atlas des pêches et des cultures marines. France, Europe, Monde*. Ouest-France-Le Marin, Paris, 1990.

Chua Th.-E., The ASEAN/US Coastal Resources Management Project: Initiation, Implementation and Management. In *Integrative Framework and Methods for Coastal Area Management*, ed. T.-E. Chua & L.F. Scura, ICLARM and US Coastal Research Management Project, Manila, 1992, pp. 71-92.

Chua Th.-E., Essential Elements of Integrated Coastal Zone Management. *Ocean and Coastal Management*, **21**(1-3), 1993, 81-108.

Churchill R.R., Fisheries issues in maritime boundary delimitation. *Marine Policy*, **17**(1), 1993, 44-57.

Cicin-Sain B., Earth Summit implementation: progress since Rio. *Marine Policy*, **20**(2), 1996, 123-143.

Cicin-Sain B., Knecht R.W. & Fisk G.W., Growth in capacity for integrated coastal management since UNCED: an international perspective. *Ocean and Coastal Management*, special issue *Earth summit implementation: progress achieved on oceans and coasts*, **29**(1-3), 1995, 93-123.

Cicin-Sain B., Multiple Use Conflicts and their Resolution: Toward a Comparative Research Agenda. In *Ocean Management in Global Change*, ed. P. Fabbri, Elsevier Applied Science, London, 1992, pp. 280-307.

Cicin-Sain B., Sustainable Development and Integrated Coastal Management. *Ocean and Coastal Management*, **21**(1-3), 1993, 11-43.

Cicin-Sain B., Knecht R.W., Measuring progress on UNCED implementation. *Ocean and Coastal Management*, special issue *Earth summit implementation: progress achieved on oceans and coasts*, 29(1-3), 1995, 1-11.

Cicin-Sain B., Knecht R.W., *Integrated Coastal and Ocean Management. Concepts and Practices*. Island Press, Washington D.C., 1998.

Clark J.R., *Coastal Zone Management Handbook*. Lewis Publishers, Boca Raton, Florida, USA, 1995.

Clark J.R., *Integrated management of coastal zones*. FAO, Rome, 1992.

Clark M.J., The realtionship between coastal-zone management and off-shore economic development. *Maritime Policy and Management*, 4(6), 1977, 431-449.

Clément E., Current developments at UNESCO concerning the protection of the underwater cultural heritage. (Presentation made at the First and the Second National Maritime Museum Conferences on the Protection of Underwater Cultural Heritage; Greenwich, 3 and 4 February 1995; London, IMO, 25 and 26 January 1996). *Marine Policy*, **20**(4), 1996, 309-323.

Cole-King A., Lalwani C.S., Information and Data Processing for Ocean Management. In *Ocean Management in Global Change*, ed. P. Fabbri, Elsevier Applied Science, London, 1992, pp. 134-152.

Cole-King A., Marine conservation: a new policy area. *Marine Policy*, **17**(3), 1993, 171-185.

Conforti B. & Francalanci G., *Atlante dei confini sottomarini /Atlas of the Seabed Boundaries*. Giuffré, Milan, Italy, 1979.

Corlay J.-P., Coastal Wetlands: A Geographical Analysis and some Projects for Management. *Ocean and Coastal Management*, **29**(1), 1993, 17-36.

Coughanowr C. & Kullenberg G., On the Need for Scientific Input on an International Scale: The Intergovernmental Oceanographic Commission's Perspective on Integrated Coastal Management. *Ocean and Coastal Management*, **21**(1/3), 1993, 339-352.

Couper A.D. (ed), *The Times Atlas of the Oceans*. Times Books, London, 1983.

Couper A.D., History of Ocean Management. In *Ocean Management in Global Change*, ed. P. Fabbri, Elsevier Applied Science, London, 1992, pp. 1-18.

Couper A.D., Sea Use Planning and Management Simulations. Ocean and Shoreline Management, **15** (4), 1991, 283-288.

Couper A.D., The principal issues in underwater cultural heritage. *Marine Policy*, **20**(4), 1996, 283-285.

Crawford B.R., Building Capacity for Integrated Coastal Management in Developing Countries. *Ocean and Coastal Management*, **21**(1/3), 1993, 311-337.

Dahl A.L., Land-based pollution and integrated coastal management. *Marine Policy*, **17**(6), 1993, 561-572.

Dejeant-Pons M., Les conventions du programme des Nations Unies pour l'environnement relatives aux mers régionales. *Annales Français de Droit International*, **33**, 1987, 690-718.

Doumenge F., La révolution aquacole. *Annales de Géographie*, **45**(530), 445-482, **45**(531), 529-586, 1986.

Edwards S.F., *An Introduction to Coastal Zone Economics: concepts, methods, and case studies*. Taylor & Francis, New York, 1987.

Edwards S.F., Estimates of Future Demographic Changes in the Coastal Zone. *Coastal Management*, **17**(3), 1989, 229-240.

Edwards S.F., What Percentage of the Population Will Live in Coastal Zones?. In *Coastal Zone '89*, ed. O.T. Magoon *et al.*, American Society of Civil Engineers, New York, Vol. I, 1989, pp. 114-126.

European Science Foundation, Commission of the European Communities, *Prediction of Change in Coastal Seas. Grand Challenges for European Cooperation in Coastal Marine Science*. ECOPS Coastal Zone

Steering Group & Ministry of Transport, Public Works and Water Management, National Institute for Coastal and Marine Management, The Hague, The Netherlands, 1993.

Falkenmark M., Main Problems of Water use and Transfer of Technology. *GeoJournal*, 3(5), 1979, 435-443.

FAO, *Atlas of the living resources of the sea*. FAO, Fishery Department, Rome, 1981.

FAO, *Integrated management of coastal zones*. Report by Clark J.R., FAO, Rome, 1992.

FAO, Problématique de la pêche mondiale. In *Gérer la mer, Cahier de la Documentation Française*, 208, 1982, 26-32.

Fenton D.M. & Syme G.J, Perception and Evaluation of the Coastal Zone: Implications for Coastal Zone Planning. *Coastal Management*, 17(3), 1989, 295-308.

Fiske S.J., Sociocultural Aspects of Establishing Marine Protected Areas. *Ocean and Coastal Management*, 17(1), 1992, 25-46.

Folke C. & Kautsky N., Aquaculture with its Environment: Prospects for Sustainability. *Ocean and Coastal Management*, 17(1), 1992, 5-24.

Ford G., Niblett C. & Walker L., *The Future for Ocean Technology*. Frances Pinter, London, 1987.

Francalanci G., Romanò & Scovazzi T., *Atlas of the Straight Baselines*, Part. I, *Art. 7 of the Convention of the United Nations on the Law of the Sea*. Giuffré, Milan, Italy, 1986.

Fricker A., Forbes D.L., A System of Coastal Description and Classification. *Coastal Management*, 16(2), 1988, 111-137.

Garcia S.M., Ocean Fisheries Management: The FAO Programme. In *Ocean Management in Global Change*, ed. P. Fabbri, Elsevier Applied Science, London, 1992, pp. 381-418.

GESAMP, *Environmental capacity: an approach to marine pollution prevention*. UNEP Regional Seas Reports and Studies, No. 80. UNPEP, Nairobi, 1986.

GESAMP, *Reducing Environmental Impacts of Coastal Aquacolture, Reports and Studies*, No. 47. FAO, Rome, 1991.

GESAMP, *The State of the marine environment*. UNEP Regional Seas Reports and Studies, No. 115. UNEP, Nairobi, 1990.

GESAMP, *Thermal discharges in the marine environment*. UNEP Regional Seas Reports and Studies, No. 45. UNEP, Nairobi, 1984.

Giordani G. & Melotti P., *Elementi di acquacoltura*. Edagricole, Bologna, Italy, 1984.

Golberg E.D., *Coastal zone space. Prelude to conflict?*. UNESCO, Paris, 1994.

Griffith M.D. & Ashe J., Sustainable Development of Coastal and Marine Areas in Small Island Developing States: A Basis for Integrated Coastal Management. *Ocean and Coastal Management*, 21(1-3), 1993, 269-284.

Guilcher A., Impact of Ocean Circulation on Regional and Global Change. In *Ocean Management in Global Change*, ed. P. Fabbri, Elsevier Applied Science, London, 1992, pp. 74-89.

Hall P., Waterfronts: a new urban frontier. *Aquapolis*, 1(1), 1992, 6-17.

Hawaii Ocean and Marine Resources Council, *Hawaii Ocean Resources Management Plan, Technical Supplement*. Hawaii Ocean and Marine Resources Council, Honolulu, 1991.

Hayashi M., Archaeological and historical objects under the United Nations Convention on the Law of the Sea. *Marine Policy*, 20(4), 1996, 291-296.

Hayashi M., The Role of National Jurisdictional Zones in Ocean Management. In *Ocean Management in Global Change*, ed. P. Fabbri, Elsevier Applied Science, London, 1992, pp. 209-226.

Herz R., Remote Sensing in Ocean Management. In *Ocean Management in Global Change*, ed. P. Fabbri, Elsevier Applied Science, London, 1992, pp. 124-133.

Holt S.J., Vanderbilt C., Marine Fisheries. In *Ocean Yearbook 2*, ed E.Mann Borgese & N. Ginsburg, University of Chicago Press, Chicago, Illinois, USA, 1980, p. 9-56.

Hoozemans F.M.J. *et al.*, *Vulnerability of coastal areas to sea-level rise: some global results*. Delft Hydraulics Publications, No. 472, Delft Hydraulics, Emmeloord, The Netherlands, 1992.

Hoozemans F.M.J., Stive M.J.F. & Bijlsma L., A Global Vulnerability Assessment: Vulnerability of Coastal Areas to Sea-Level Rise. In *Coastal Zone '93*, ed. T.O. Magoon *et al.*, American Society of Civil Engineers, New York, Vol. I, 1993, pp. 390-404.

Hoyle B.S. & Pinder D.A., Husain M.S. ed, *Revitalising the Waterfront*. Belhaven Press, London, 1988.

Hoyle B.S., Development dynamics at the port-city interface. In *Revitalising the waterfront. International dimensions of dockland redevelopment*, ed. B.S. Hoyle B.S., D.A. Pinder & M.S. Husain, Belhaven, London, 1988, pp. 3-19.

Hufschmidt M.M., A conceptual framework for watershed management. In *Watershed resources management: an integrated framework with studies from Asia and the Pacific*, ed. K.W. Easter, J.A. Dixon & M.M. Hufschmidt, Westview Press, Boulder, Colorado, USA, 1986.

IMO, FAO, UNESCO, WMO, IAEA, UN, UNEP, *Coastal Modelling*. GESAMP Reports and Studies, No. 43, IAEA, Wien, 1991.

IMO, FAO, UNESCO,WMO, WHO, IAEA, UN, UNEP, *Global Strategies for Marine Environmental Protection*. GESAMP Reports and Studies, No. 45. IMO, London, 1991.

Intergovernmental Oceanographic Commission, *Ocean science for the year 2000*. UNESCO, Paris, 1984.

Intergovernmental Panel on Climate Change (IPCC), *Preparing to meet the Coastal Challenges of the 21st centiry. Conference Report, World Coast Conference 1993*, Ministry of Transport, Public Works and Water Management, National Institute for Coastal and Marine Management, Coastal Zone Management Centre, The Hague, The Netherlands, 1994.

Intergovernmental Panel on Climate Change, *Global Climate Change and the Rising Challenge of the Sea*, ed L. Bijlsma *et al.*, Ministry of Transport, Public Works and Water Management, The Hague, The Nethrlands, 1992.

International Geosphere-Biosphere Programme, *The International Geosphere-Biosphere Programme: A study of global change IGBP. A plan for action*, Global Change, Report No. 4, IGBP, Stockholm, 1988.

International Geosphere-Biosphere Programme, *Land-Ocean Interactions in the Coastal Zone. Science Plan*. Global Change, Report No. 25, IGBP, Stockholm, 1993.

International Geosphere-Biosphere Programme, *Land-Ocean Interactions in the Coastal Zone. Implementation Plan*. Global Change, Report No. 33, IGBP, Stockholm, 1994.

Jelgersma S., Tooley M.J., Impacts of a Future Sea-level Rise on European Coastal Lowlands. In *Impacts of Sea-level Rise on European Coastal Lowlands*, ed. M.J. Tooley & S. Jelgersma, Blackwell, Oxford/Cambridge, UK, 1992, pp. 1-35.

Jelgersma S., Van Der Zijp M. & Brinkman R., Sealevel Rise and the Coastal Lowlands in the Developing World. *Journal of Coastal Research*, 9(4), 958-972.

Johnson J.C. & Pollnac R.B., Introduction to Managing Marine Conflicts. *Ocean and Shoreline Management*, 12(3), 1989, 191-198.

Jolliffee I.P. & Patman C.R., The Coastal Zone: The Challenge, *Journal of Shoreline Management*, 1(1-2), 1985, 3-36.

Kam S.P., Paw J.N. & Loo M., The Use of Remote Sensing and Geographic Information Systems in Coastal Zone Management. In *Integrative Framework and Methods for Coastal Area Management*, ed. T-.E. Chua & L.F. Scura, ICLARM and US Coastal Resource Management Project, Manila, 1992, pp. 107-131.

Kapetsky J.M. & Lasserre G., *Management of coastal lagoon fisheries*. FAO, Rome, 1984.

Keckes S., The Protection and Development of the Marine Environment: UNEP's Oceans and Coastal Areas Programme. In *Ocean Management in Global Change*, ed. P. Fabbri, Elsevier Applied Science, London, 1992, pp. 344-360.

Kenchington R., *Managing Marine Environments*. Taylor & Francis, New York, USA, 1990.

Kenchington R. & Crawford D., On the meaning of integration in coastal zone management. *Ocean and Coastal Management*, 21(1-3), 1993, 109-127.

Kennett J., *Marine Geology*. Prentice-Hall, Englewood Cliffs, UK, 1982.

King J., Beyond the shoreline. *Marine Policy*, 18(6), 1994, 457-463.

King L.R. & Olson S.G., Coastal State Capacity for Marine Resources Management. *Coastal Management*, 16(4), 1988, 305-318.

Laughlin T.L. & Sherman K., Large Marine Ecosystems: A New Concept in Ocean Management. In *Coastal Ocean Space Utilization III*, ed. N. Della Croce N., S. Connell & R. Abel R., E.&FN Spon, London, 1995, pp. 283-293.

Le Moigne J.-L., *La modélisation des systèmes complexes*. Dunod, Paris, 1990.

Le Moigne J.-L., *La théorie du système général. Théorie de la modélisation.* Presses Universitaires de France, Paris, 1977-1994.

Mann Borgese E., *The Future of the Oceans.* Harvest House, Montreal, Canada, 1986.

Miles E.L., Future Challenges in Ocean Management: Towards Integrated National Ocean Policy. In *Ocean Management in Global Change*, ed. P. Fabbri, Elsevier Applied Science, London, 1992, pp. 595-620.

Miller M.L. & Kirk J., Marine Environmental Ethics. *Ocean and Coastal Management*, **17**(3), 1992, 237-250.

Miller M.L., The Rise of Coastal and Marine Tourism. *Ocean and Coastal Management*, **20**(3), 1993, 181-200.

Milliman J.D., Management of the Coastal Zone: Impact of Onshore Activities on the Coastal Environment. In *Use and Misuse of the Seafloor*, ed. K.J. Hsu & J. Thiede, Wiley & Sons, New York, 1992, pp. 213-227.

Mitchell J.K., Coastal Zone Management: A Comparative Analysis of National Programs. In *Ocean Yearbook* 3, ed E. Mann Borgese & N. Ginsburg, University of Chicago Press, Chicago, Illinois, USA, 1982, 258-319.

Morgan J.R., The Marine Region. *Ocean and Coastal Management*, **24**(1), 1994, 51-70.

Morin E., *La métode*, Vol. III *La Connaissance de la Connaissance/1: Anthropologie de la Connaissance.* Seuil, Paris, 1986.

Morin E., *La métode*; Vol. I *La Nature de la Nature.* Seuil, Paris, 1977.

Mörner N.-A., Models of global sea-level changes. In *Sea-level changes*, ed. M.J. Tooley & I. Shennan, Blackwell, Oxford, UK, 1987, pp. 332-355.

Morris M.A., Military Aspects of the Exclusive Economic Zone. In *Ocean Yearbook 3*, ed E. Mann Borgese & N. Ginsburg, University of Chicago Press, Chicago, Illinois, USA, 1982, p. 320-348.

National Economic Development Authority (Region 1, Philippines), *The Lingayen Gulf coastal area management plan.* International Center for Living Resources Management, Technical Reports, No. 32, Manila, 1992.

National Research Council, *Priorities for Coastal Ecosystem Science.* National Academy Press, Washington D.C., 1994.

Nicholls R.J. & Leatherman S.P., Sea-Level Rise and Coastal Management. In *Geomorphology and Land Management in a Changing Environment*, ed. D.F.N. McGregor & D.A. Thompson, Wiley & Sons, New York, 1995, pp. 229-244.

Nicholls R.J. & Leatherman S.P., The Implications of Accelerated Sea-Level Rise for Developing Countries: A Discussion. *Journal of Coastal Research*, special issue n. 14 (R.J. Nicholls & S.P. Leatherman ed), 1995, 303-323.

Nicholls R.J., Leatherman S.P., Dennis K.C. & Volonté C.R., Impacts and Responses to Sea-Level Rise: Qualitative and Quantitative Assessments. *Journal of Coastal Research*, special issue n. 14 (R.J. Nicholls & S.P. Leatherman ed), 1995, 26-43.

Nonn H., *Géographie des littoraux.* Presses Universitaires de France, Paris, 1972.

Nordquist M.H. (ed), *United Nations Convention on the Law of the Sea 1982. A Commentary*, vol.I. Martinus Nijhoff Publications, Dordrecht, The Netherlands, 1985.

Norse E. A. (ed), *Global Marine Biological Diversity*, Island Press, Washington D.C., 1993.

ODUM E.P. ,*Fundamentals of ecology.* Philadelphia, Saunders, 1971.

OECD, *Coastal zone management. Integrated Policies*, OECD, Paris, 1993. *Quoted as OECD/a.*

OECD, *Coastal zone management. Selected case studies*, OECD, Paris, 1993. *Quoted as OECD/b.*

Olsen S., The skills, knowledge, and attitudes of an ideal coastal manager. In *Educating coastal managers: proceedings of the Rhode Island workshop* , ed. B. R. Crawford, J. S. Cobb & C. L. Ming, Coastal Resources Center, Rhode Island, USDA, 1995, 3-7.

Orford J., Coastal Processes: The Coastal Response to Sea-level Variation. In *Sea surface studies*, ed. R.J.N. Devoy, Croom Helm, London-New York-Sydney, 1987, pp. 415-463.

Paskoff R., *Les littoraux. Impact des aménagement sur leur évolution*, Masson, Paris, 1985.

Peck D.L. & Williams S.J., Sea-Level Rise and its Implications in Coastal Planning and Management. In *Ocean Management in Global Change*, ed. P. Fabbri, Elsevier Applied Science, London, 1992, pp. 57-73.

Peet G., Ocean Management in Practice. In *Ocean Management in Global Change*, ed. P. Fabbri, Elsevier Applied Science, London, 1992, pp. 39-56.

Pernetta J.C. & Elder D.L., Climate, Sea Level Rise and the Coastal Zone: Management and Planning for Global Changes. *Ocean and Coastal Management*, **18**(1), 1992, 113-160.

Pernetta J.C., Impacts of climate change and sea-level rise on small island states. *Global Environmental Change*, **2**(1), 1992, 19-31.

Pethick J., *An Introduction to Coastal Geomorphology*, Arnold, London, 1988.

Pido M.D. and Chua T.-E., 'A Framework for Rapid Appraisal of Coastal Environments', In: Chua T.-E. and Fallon Scura L. (eds) *Integrative Framework and Methods for Coastal Area Management*. International Center for Living Resources Management, Manila, 1992, pp. 144-7.

Pinder D.A. & Hoyle B.S., Urban Waterfront Management: Historical Patterns and Prospects. In *Ocean Management in Global Change*, ed. P. Fabbri, Elsevier Applied Science, London, 1992, pp. 482-501.

Prescott J.V.R., Delimitation of Marine Boundaries by Baseline. *Marine Policy Reports*, **8**(3), 1986, 1-5.

Prescott J.V.R., *The Maritime Political Boundaries of the World*, Methuen, London, 1985.

Prescott V., Boundaries and Ocean Management. In *Ocean Management in Global Change*, ed. P. Fabbri, Elsevier Applied Science, London, 1992, pp. 227-246.

Psuty N.P., Estuaries: Challenges for Coastal Management. In *Ocean Management in Global Change*, ed. P. Fabbri, Elsevier Applied Science, London, 1992, pp. 502-520.

Ricklefs R.E., Scaling Pattern and Process in Marine Ecosystems", in *Large Marine Ecosystems: Patterns, Processes and Yields*, ed. K. Sherman, L.M. Alexander & B.D. Gold, American Association for Advancement Science, Washington D.C., 1990, pp. 169-178.

Robinson N.A. (ed), *Agenda 21 and UNCED Proceedings*. Oceana, New York, USA, No. 6 volumes, 1992.

Salm R.V. & Clark J.R., *Marine and Coastal Protected Areas: A Guide For Planners and Managers*, International Union for Conservation of Nature and Natural Resources, Gland, France, 1989.

Schmidheiny S. & Business Council for Sustainable Development (1992): *Changing Course. A Global Business Perspective on Development and the Environment*. MIT Press, Cambridge, Massachusetts, USA, 1992.

Schröder P.C., The Need for International Training in Coastal Management. *Ocean and Coastal Management*, **21**(1-3), 1993, 303-310.

Schröder P.C., UNEP's Regional Seas Programme and the UNCED Future: Aprés Rio. *Ocean and Coastal Management*, **18**(1), 1992, 101-112.

Scovazzi T., Francalanci G., Romanò D., Mongardini S., *Atlas of the Straight Baselines*, Giuffré, Milan, Italy, 2nd ed., 1989.

Scura L.F., Chua T.-E., Pido M.D. & Paw J.N., Lessons for Integrated Coastal Zone Management: The ASEAN Experience. In *Integrative Framework and Methods for Coastal Area Management*, ed. T-.E.Chua & L.F. Scura, ICLARM and US Coastal Reseurce Management Project, Manila, 1992, pp. 1-70.

Shaw J. & Forbes D.L., Relative Sea-Level Change and Coastal Response, Northeast Newfoundland. *Journal of Coastal Research*, **6**(3), 1990, 641-660.

Shennan I. & Tooley M.J., Conspectus of fundamental and strategic research on sea-level changes. In *Sea-level changes*, ed. M.J. Tooley & I. Shennan, Blackwell, Oxford, UK, 1987, pp. 371-390.

Sherman K., Measurement strategies for monitoring and forecasting variability in large marine ecosystems. In *Variability and Management of Large Marine Ecosystems*, ed. K. Sherman & L.M. Alexander, Westview Press, Boulder, Colorado, USA, 1986, pp. 203-236.

Sherman K. & Alexander L.M. ed, *Variability and Management of Large Marine Ecosystems*, Westview Press, Boulder, Colorado, USA, 1986.

Smith H.D., The application of maritime geography: a technical and general management approach. In *The Development of Integrated Sea-Use Management*, ed. H.D. Smith & A. Vallega, Routledge, London, 1991, pp. 7-16.

Smith H.D., The Development and Management of the World Ocean. *Ocean and Coastal Management* **24**(1), 1994, 3-16.

Smith H.D., The Regional Bases of Sea Use Management. *Ocean and Shoreline Management*, **15**(4), 1991, 273-282.

Smith H.D., The role of the state in the technical and general management of the oceans. *Ocean and Coastal Management*, **27**(1-2), 1995, 5-14.

Smith H.D., Theory of Ocean Management. In *Ocean Management in Global Change*, ed. P. Fabbri, Elsevier Applied Science, London, 1992, pp. 19-38.

Smith R.W., Roach J.A., *Limits in the Seas*, United States Department of State/Bureau of Oceans and International Environmental and Scientific Affairs, Washington, D.C., 1992.

Sorensen J., The International Proliferation of Integrated Coastal Zone Management Efforts. *Ocean and Coastal Management*, **21**(1-3), 1993, 45-80.

Sorensen J., The management of enclosed coastal water bodies: the need for a framework for international information exchange. In *The management of coastal lagoons and enclosed bays*, ed. J. Sorensen, F. Gable & F Bandarin, American Society of Civil Engineers, New York, 1993, pp. 1-17.

Sorensen J.C., National and International Efforts at Integrated Coastal Management: Definitions, Achievements, and Lessons. *Coastal management*, **25**, 1997, 3-41.

Sorensen J.C., McCreary S.T., *Institutional Arrangements for Managing Coastal Resources and Environments*, U.S. National Park Service, Washington D.C., 1990.

Sorensen J., West N., *A Guide to Impact Assessment in Coastal Environments*, Coastal Resources Center and the University of Rhode Island, Rhode Island, USA, 1992.

Stewart J.R., Marine Ecumenopolis. *Ekistics*, (29), 175, 1970, 399-418.

Tait R.V., *Elements of Marine Ecology. An Introductory Course*, Butterworths, London, 1983.

The World Bank, Environment Department, *Guidelines for integrated coastal zone management. Land, water and natural habitats division*, The World Bank, 1993 (manuscript proceedings of a technical workshop, held in 1992).

The World Commission on Environment and Development, *Our common future*. Oxford University Press, Oxford, UK, 1987.

Throckmorton P. (ed), *Atlante di archeologia subacquea*. Istituto Geografico De Agostini, Novara, Italy, 1988.

Turner R.E. & Rabalais N.N., Eutrophication and its Effects on Coastal Habitats. In *Coastal Wetlands*, ed. O.T. Magoon. & H.S. Bolton, American Society of Civil Engineers, New York, 1991, pp. 61-74.

Uderdahl A., Integrated Marine Policy: What? Why How?. *Marine Policy*, July 1980, 159-169.

UNEP, *Achievements and planned development of UNEP's Regional Seas Programme and comparable programmes sponsored by other bodies*. UNEP Regional Seas Reports and Studies, No. 1. UNEP, Nairobi, 1982.

UNEP, *Guidelines and principles for the preparation and implementation of comprehensive action plans for the protection and development of marine and coastal areas of regional sea*. UNEP Regional Reports and Studies No. 15, UNEP, Nairobi, 1982.

UNEP (1995) *Guidelines for Integrated Management of Coastal and Marine Areas - with Special Reference to the Mediterranean Basin*. UNEP Regional Seas Reports and Studies, No. 161. Split, PAP/RAC (MAP-UNEP), 1995.

UNEP, *The health of the oceans*. UNEP Regional Seas Reports and Studies No. 16, UNEP, Nairobi, 1982.

United Nations Secretary-General, Coastal Area Development and Management and Marine and Coastal Technology. In *Ocean Yearbook 1*, ed E. Mann Borgese & N. Ginsburg, University of Chicago Press, Chicago, Illinois, USA, 1978, p. 350-375. This article reproduces the UN Document E/5971.

United Nations, *Coastal Area. Management and Development*. Pergamon Press, Oxford, UK, 1982.

Vallega A., A Conceptual Approach to Integrated Coastal Management. *Ocean and Coastal Management*, **21**(1-3), 1993, 149-162.

Vallega A., *Governo del mare e sviluppo sostenibile. Conoscenze di base*. Mursia, Milan, Italy, 1993.

Vallega A., *La regione, sistema territoriale sostenibile. Compendio di geografia regionale sistematica.* Mursia, Milan, Italy, 1995.

Vallega A., *Ocean change in global change. Introductory geographical analysis.* Università degli Studi di Genova and International Geographical Union, Genoa, Italy, 1990.

Vallega A., *Sea management. A theoretical approach.* Elservier Applied Science, London, 1992.

Vallega A., *The Changing Waterfront in Coastal Area Management.* Milano, Franco Angeli, 1992.

Vallega A., The Coastal Use Framework as a Methodological Tool for Coastal Area Management. In *Coastal Ocean Space Utilization III*, ed. N. Della Croce, S. Connell & R. Abel, E.&FN Spon, London, 1995, pp. 117-130.

Vallega A., Geographical coverage and effectiveness of the UNEP Convention on the Mediterranean. *Coastal and Ocean Management*, special issue *Sustainable development ot the regional level: the Mediterranea* (ed. A. Vallega), **31**(2-3), 1996, 199-218.

Vallega A., The coastal use structure within the coastal system. A sustainable development-consistent approach, in *Journal of Marine Systems*, **7**, 1996, 95-115.

Vallejo S. M., Integrated Marine Policies: Goals and Constraints. In *Ocean Management in Global Change*, ed. P. Fabbri, Elsevier Applied Science, London, 1992, pp. 153-168.

Vallejo S.M., Development and Management of Coastal and Maritime Areas: An International Prospective. In *Ocean Yearbook 7*, ed E. Mann Borgese & N. Ginsburg, University of Chicago Press, Chicago, Illinois, USA, 1988, p. 205-222.

Vallejo S.M., The Integration of Coastal Zone Management into National Development Planning. *Ocean and Coastal Management*, **21**(1-3), 1993, 163-182.

Van Der Weide J., A Systems View of Integrated Coastal Management. *Ocean and Coastal Management*, **21**(1-3), 1993, 129-148.

Van Der Weide J., Climate Change and Coastal Vulnerability. In *Coastal Ocean Space Utilization III*, ed. N. Della Croce, S. Connell & R. Abel, E.&FN Spon, London, 1995, pp. 351-367.

Van Der Weide J., Integrated Coastal Management: Past, Present and Future. In *Regional seas towards sustainable development*, ed. S. Belfiore, M.G. Lucia & E. Pesaro, Franco Angeli, Milan, Italy, 1996, pp. 157-167.

Vellinga P. & Klein R.J.T., Climate Change, Sea Level Rise and Integrated Coastal Zone Management: An IPCC Approach. *Ocean and Coastal Management*, **21**(1-3), 1993, 245-268.

Vigarié A., Maritime Industrial Development Areas: Structural Evolution and Implications for Regional Development. In *Cityport Industrialization and Regional Development. Spatial Analysis and Planning Strategies*, ed. B.S. Hoyle & D.A. Pinder, Pergamon Press, Oxford, 1981, 23-37.

Vigarié A., Ocean Sciences and Management. In *Ocean Management in Global Change*, ed. P. Fabbri, Elsevier Applied Science, London, 1992, pp. 108-123.

Walker H.J., Sea Level Change. Environmental and Socio-Economic Impacts. *GeoJournal*, **24**(4), 1992, 511-520.

Walker H.J., The Coastal Zone. In *The earth as transformed by human action. Global and regional changes in the biosphere over the past 300 years*, ed. B.L. Turner, Cambridge University Press, Cambridge, UK, 1990, pp. 271-294.

Willmann R., Insull D., Integrated Coastal Fisheries Management. *Ocean and Coastal Management*, 21(1-3), 1993, 285-302.

Wilson J.A., Townsend R., Kelban P., McKay S. & French J., Managing Unpredictable Resources: Traditional Policies Applied to Chaotic Populations. *Ocean and Shoreline Management*, **13**(3), 1990, 179-198.

Wonham J., Special Areas and Particularly Sensitive Areas.In *Ocean Management in Global Change*, ed. P. Fabbri, Elsevier Applied Science, London, 1992, pp. 361-380.

Young M.D., *Sustainable Investment and Resource Use. Equity, Environmenmtal Integrity and Economic Efficiency.* Man and Biosphere Series, Vol. 9, UNESCO and Parthenon Publishing Group, Paris, 1992.

The Mediterranean

Abdel-Aal F.M., "Sedimentation of Nile Delta Coast", in *Coastal Zone '89*, ed. O.T. Magoon *et al.*, American Society of Civil Engineers, New York, 4, 1989, pp. 4321-4333.

Abel R.B., Cooperative Mediterranean oceanography in the post Aswan Dam era. In *Symposium Mediterranean Seas 2000*, ed. N. Della Croce, Istituto di Scienze Ambientali Marine, Università di Genova, S. Margherita Ligure, Italy, 1993, pp. 327-350.

Al-Shalabei N.M. *et al.*, Implications of Expected Climatic Changes for the Syrian Coast. In *Climatic Change and the Mediterranean*, vol. 2, ed. L. Jeftic, S. Keckes & J.C. Pernetta, Arnold, London, 1996, pp. 251-321.

Amanieu M. & Lasserre G., Niveaux de production des lagunes littorales méditerranéennes et contribution des lagunes à l'enrichissement des pêches démersales. In *Management of living resources in the Mediterranean coastal area*, FAO, Rome, 1981, pp. 75-87.

Aschmann H., Distribution and Peculiarity of Mediterranean Ecosystems. In *Mediterranean Type Ecosistems*, ed. F. Di Castri & H.A. Mooney, Springer-Verlag, Berlin, 1973, pp. 11-20.

Aschmann H., Man's Impact on the Several Regions with Mediterranean Climates. In *Mediterranean Type Ecosystems*, ed. F. Di Castri & H.A. Mooney, Springer-Verlag, Berlin, 1973, pp. 363-372.

Attard D.J., The Delimitation of the Continental Shelf and the Exclusive Economic Zone in the Mediterranean Sea. In *Il regime giuridico internazionale del Mare Mediterraneo*, ed. U. Leanza, Giuffré, Milan, Italy, pp. 77-80, 1987.

Bahri-Maddeb A., Binous J., Drocourt D., Jokilehto J., Marasovic J., Marasovic T., Morozzo M.D., Zivas D., *Guidelines for the rehabilitation of Mediterranean historical settlments*, 2 volumes. MAP, Priority Actions Programme, Split, Croatia, 1994.

Balkas T.I., Yetis Ü. & Tuncel G., Coastal zone management studies in relation to Mediterranean Action Plan: Izmir Bay case study. In *Symposium Mediterranean Seas 2000*, ed. N. Della Croce, Istituto di Scienze Ambientali Marine, Università di Genova, S. Margherita, Italy, 1993, pp. 289-306.

Bandarin F., The management of coastal lagoons: the approach of the Venice safeguard project. In *The management of coastal lagoons and enclosed bays*, ed J. Sorensen, F. Gable & F. Bandarin, New York, American Society of Civil Engineers, 1993 94-109.

Bandarin F., The small islands of the Mediterranean: Development issues and environmental management. In *Proceedings of the Second International Conference on the Mediterranean Coastal Environment medcoast'95*, ed. E. Özhan E., Middle East Technical University, Ankarta, vol. 1, 1995, pp. 537-536.

Baric A. & Gasparovic F., Implications of Climatic Change on the Socio-Economic Activities in the Mediterranean Coastal Zones. In *Climatic Change and the Mediterranean*, vol. 1, ed. L. Jeftic, L.D. Milliman & G. Sestini, Arnold, London, 1992, pp. 129-174.

Baric A. *et al.*, Implications of Expected Climatic Changes for the Kastela Bay Region of Croatia. In *Climatic Change and the Mediterranean*, vol. 2, ed. L. Jeftic, S. Keckes & J.C. Pernetta, Arnold, London, 1996, pp. 143-249.

Bellan-Santini D. & Leveau M., Case Study: Eutrophication in the Golfe du Lion. In *Eutrophication in the Mediterranean Sea: deceiving capacity and monitoring of long term effects*, Athens, UNEP, 1988, pp. 107-22.

Bethoux J.P. & Gentili B., The Mediterranean Sea, coastal and deep-sea signatures of climatic and environmental changes. *Journal of Marine Systems*, 7(2-4), 1996, 383-394.

Bethoux J.P., The Mediterranean Sea, a biogeochemistry model marked by climate and environment. In *Symposium Mediterranean Seas 2000*, ed. N. Della Croce, Istituto di Scienze Ambientali Marine, Università di Genova, S. Margherita, Italy, 1993, pp. 351-371.

Blake G.H., Mediterranean Micro-Territorial Disputes and Maritime Boundary Delimitation. In *Il regime giuridico internazionale del Mare Mediterraneo*, ed. U. Leanza, Giuffré, Milan, Italy, 1987, pp. 111-118.

Bondesan M. *et al.*, Coastal Areas at Risk from Storm Surges and Sea-Level Rise in Northeastern Italy. *Journal of Coastal Research*, 11(4), 1995, 1354-1379.

Bouhlel M., Aménagement des ressources vivantes de la zone littorale tunisienne. In *Management of living resources in the Mediterranean coastal area*, FAO, Rome, 1981, pp. 95-130.

Bowett D.W., Exploitation of Mineral Resources and the Continental Shelf. In *Il regime giuridico internazionale del Mare Mediterraneo*, ed. U. Leanza, Giuffré, Milan, Italy, 1987, pp. 21-32.

Brambati A., Sedimentology and Pollution in the Mediterranean: a Discussion. In *The Mediterranean Sea*, ed. D.J. Stanley, Hutchinson & Ross, Strasbourg, France, 1972, pp. 711-721.

Brambati A., Some Aspects of the Mineral Resources of the Mediterranean Sea. *Lo spettatore internazionale*, **17**(4), 1982, 307-317.

Brigand L. *et al.*, *Les îles en Méditerranée. Enjeux et perspectives*. Economica, Paris, 1991.

Briguglio L. & Briguglio M., The economic and environmental impacts of tourism on the Maltese Islands. In *Proceedings of the Second International Conference on the Mediterranean Coastal Environment Medcoast'95*, ed. E. Özhan, Middle East Technical University, Ankara, vol. 1, 1995, pp. 321-334.

Broadus J.M. *et al.*, *Comparative Assessment of Regional International Programs to Control Land-Based Marine Pollution: The Baltic, North Sea, and Mediterranean*. Woods Hole Oceanographic Institution, Woods Hole, Massachusetts, USA, 1993.

Bruckner H., Man's Impact on the Evolution of the Physical Environment in the Mediterranean Region in Historical Times. *GeoJournal*, **13**(1), 1986, 7-17.

Brunel G. *et al.*, Aquaculture and coastal planning in the Mediterranean. In *Environmental aspects of aquaculture in the Mediterranean region. Documents produced in the period 1985-1987*, MAP Technical Reports Series No. 15, UNEP, Priority Actions Programme, Regional Activity Centre, Split, Croatia, 1987, pp. 3-32.

Busuttil S., The Future of the Mediterranean. In *Ocean Yearbook 10*, ed E. Mann Borgese, N. Ginsburg & J.R. Morgan, University of Chicago Press, Chicago, Illinois, USA, 1993, p. 240-247.

Caddy J.F. & Griffiths R.C., *Tendances recentes des pêches et de l'environnement dans la zone couverte per le Conseil Général des pêches pour la Méditerranée*. FAO, Études et Revues, No. 63, FAO, Rome, 1991.

Caddy J.F., Refk R. & Do-Chi T., Productivity estimates for the Mediterranean: evidence of accelerating ecological change. *Ocean and Coastal Management*, **26**(1), 1995, 1-18.

Calcagnile G. & Panza G.F., The Main Characteristics of the Lithosphere-Astenosphere System in Italy and Surrounding Regions. *Paleogeography*, 119, 1980, 865-879.

Calvo Garcia Tornel F., L'arc méditerranéen espagnol, un espace désarticulé. *Méditerranée*, 79(1-2), 1994, 51-60.

Cavallaro C., *Sfruttamento e utilizzazione delle fonti alternative di energia nelle Isole Eolie*. Programma UNESCO-MAB, Sagep, Genoa, Italy, 1986.

Charbonnier D. et al., *Pêche & aquaculture en Méditerranée. Etat actuel et perspectives*. Economica, Paris, 1990.

Chiappori A., Balostro R., Thake B., Santillo D., Thornton D. & Patel D., The Adriatic Sea and Coastal Resources: A Management and Pollution Control Study. In *Coastal Ocean Space Utilization III*, ed. N. Della Croce, S. Connell & R. Abel, E.&FN Spon, London, 1995, pp. 331-344.

Chircop A., *The Law of the Sea and the Mediterranean*. UNEP, Nairobi, 1991.

Chircop A.E., The Mediterranean Sea and the Quest for Sustainable Development. *Ocean Development and International Law*, 23, 1992, 17-30.

Cicin-Sain B., Sustainable development, integrated coastal management and tourism: Challenges to Mediterranean Countries, in *Medcoast 93*, v.1, ed. E. Özhan, Medcoast Institute, Ankara, 1993, pp. 13-34.

Cognetti G. & C. Da Pozzo, Marine parks in the Mediterranean. In *The Development of Integrated Sea-Use Management*, ed. H.D. Smith & A. Vallega, Routledge, London, 1991, pp. 186-196.

Conforti B., The Mediterranean and Exclusive Economic Zone. In *Il regime giuridico internazionale del Mare Mediterraneo*, ed. U. Leanza, Giuffré, Milan, Italy, 1987, pp. 173-182.

Cori B. & Cortesi G., Systems of town and systems of ports: the case of Italy. In *Urban Systems in Transition*, ed. J.G. Borchert, L.S. Bourne & R. Sinclair, Department of Geography, Utrecht, The Netherlands, 1986, pp. 134-141.

Demetropoulos A., The management and control of coastal fisheries. In *Management of living resources in the Mediterranean coastal area*, FAO, Rome, 1981, pp. 23-34.

Di Castri F. & Mooney A.H., *Mediterranean type ecosystem. Origin and structure*. Springer-Verlag, Heildelberg, Germany, 1973.

Dobbin J.A.& Trumbic I., Planning for the future of the national coast: The Albanian coastal area management project. In *Proceedings of the Second International Conference on the Mediterranean Coastal Environment MEDCOAST'95*, ed. E. Õzhan, Middle East Technical University, Ankara, vol. 1, 1995, pp. 391-415.

Dounas A., Evaluation of the water reserves and operation of the karstic aquifer. In *Water resources development of small Mediterranean islands and isolated coastal areas. Documents produced in the first stage of the Priority Action (1984-1985)*. MAP Technical Reports Series No. 12, UNEP, Priority Actions Programme, Regional Activity Centre, Split, Croatia, 1987, pp. 39-62.

Drakopoulos J., Greek national report concerning land-use in seismic zones. In *Land use planning in earthquake zones. Seismic risk reduction in the Mediterranean region. Documents produced in the first stage of the Priority Action (1984-1986)*, MAP Technical Reports Series No. 17, UNEP, Priority Actions Programme, Regional Activity Centre, Split, Croatia, 1987, pp. 21-36.

Dufournet Y. & Saliba L., Advanced wastewater treatement for industrial reuse. In *Water resources development of small Mediterranean islands and isolated coastal areas. Documents produced in the first stage of the Priority Action (1984-1985)*, MAP Technical Reports Series No. 12, UNEP, Priority Actions Programme, Regional Activity Centre, Split, Croatia, 1987, pp. 86-103.

El Sayed M.K., Implications of Relative Sea Level Rise on Alexandria. In *Impact of sea level rise on cities and regions*, Venice, First International Meeting 'Cities on Water', ed. R. Frassetto, Marsilio, Padova, Italy, 1991, pp. 183-9.

El-Fishawi N.M. & A.A. Badr, Erosion and flooding as related to recent sea level rise, Nile Delta coast, Egypt. In *Proceedings of the Second International Conference on the Mediterranean Coastal Environment MEDCOAST'95*, ed. E. Õzhan, Middle East Technical University, Ankara, vol. 2, 1995, pp. 1229-1239.

Eleftheriou A. & Smith C.J., Preliminary investigations of the benthic ecosystem from the aegean shelf (Eastern Mediterranean). In *Symposium Mediterranean Seas 2000*, ed. N. Della Croce, Istituto di Scienze Ambientali Marine, Università di Genova, S.Margherita Ligure, Italy, 1993, pp. 105-120.

Emeis K.C., Larsen B. & Seibold E., What Is the Environmental Capacity of Enclosed Marginal Seas? Approaches to the Problem in the Baltic, North, and Mediterranean Seas. In *Use and Misuse of the Seafloor*, ed. K.J. Hsu & J. Thiede, Wiley & Sons, New York, 1992, 181-211.

Ergunay O., Earthquake hazard and mitigation in Turkey. In *Land use planning in earthquake zones. Seismic risk reduction in the Mediterranean region. Documents produced in the first stage of the Priority Action (1984-1986)*, MAP Technical Reports Series No. 17, UNEP, Priority Actions Programme, Regional Activity Centre, Split, Croatia, 1987, pp. 37-55.

Esmer Õ., Transforming the Mediterranean Blue Plan into a system dynamics model. In *Medcoast 93*, v.1, ed. E. Õzhan, Medcoast Institute, Ankara, 1993, pp. 273-290.

Fabbri P., Management of the sea: the case of the Mediterranean. In *The new frontiers of marine geography/Les nouvelles frontières de la géographie de la mer*, ed. H.D. Smith & A. Vigarié A., 1988, pp. 99-103.

FAO, *Projet de développement méditerranéen. Développement agro-sylvicole et expansion économique dans les pays méditerranéens. Plan d'action pour un développement intégré*. FAO, Rome, 1959.

Flemming N.C., Predictions of Relative Coastal Sea-Level Change in the Mediterranean Based on Archaeological, Historical and Tide-Gauge Data. In *Climatic Change and the Mediterranean*, vol. 1, ed. L. Jeftic, L.D. Milliman & G. Sestini., Arnold, London, 1992, pp. 247-281.

Frihy O.E., Some proposals for coastal management of the Nile delta coast. *Ocean and Coastal Management*, **30**(1), 1996, 43-59.

Gacic M., Hopkins T.S. & LASCARATOS A., Aspects of the Response of the Mediterranean Sea to Long-Term Trends in Atmospheric Forcing. In *Climatic Change and the Mediterranean*, vol. 1, ed. L. Jeftic, L.D. Milliman & G. Sestini, Arnold, London, 1992, pp. 233-246.

Garcia A., Crespo J. & Rey J.C., A contribution to knowledge of the southern part of Spain's Mediterranean coastal zone including a description of a beach seine fishery. In *Management of living resources in the Mediterranean coastal area*, FAO, Rome, 1981, 131-147.

Gatescu P., The Danube Delta: Geographical Characteristics and Ecological Recovery. *GeoJournal*, 29(1), 1993, 57-67.

Giri J., *Industrie et environnement en Méditerranée. Evolution et perspectives*. Economica, Paris, 1991.

Grenon M. & Batisse M., *Futures for the Mediterranean Basin. The Blue Plan*. Oxford University Press, London, 1989.

Grenon M. *et al.*, *Energie et environnement en Méditerranée. Enjeux et prospective*. Economica, Paris, 1993.

Grofit E., The artisanal coastal fishery in the Eastern Mediterranean. In *Management of living resources in the Mediterranean coastal area*, FAO, Rome, 1981, 1-22.

Haas P.M., Save the Seas: UNEP's Regional Seas Programme and the Coordination of Regional Pollution Control Efforts. In *Marine Policy Yearbook*, Chicago University Press, Chicago, Illinois, USA, 1991, pp. 188-212.

Haas P.M., *Saving the Mediterranean. The Politics of International Environmental Cooperation*. Columbia University Press, New York, 1990.

Holdsworth B., Sustainable tourism. In *Medcoast 93*, v.1, ed. E. Özhan E., Medcoast Institute, Ankara, 1993, pp. 229-242.

Hollis G.E, Implications of Climatic Changes in the Mediterranean Basin. In *Climatic Change and the Mediterranean*, vol. 1, ed. L. Jeftic, L.D. Milliman & G. Sestini, Arnold, London, 1992, pp. 602-665.

Imeson A.C. & Emmer I.M., Implications of Climatic Change on Land Degradation in the Mediterranean. In *Climatic Change and the Mediterranean*, vol. 1, ed. L. Jeftic, L.D. Milliman & G. Sestini G., Arnold, London, 1992, pp. 95-128.

Jagota S.P., The Law of the Sea. A Historical Perspective. In *The Mediterranean in the New Law of the Sea*, Foundation for International Studies, Valletta, 1987, pp. 25-32.

Jeftic L., Integrated coastal and marine areas management (ICAM) in the Mediterranean Action Plan of UNEP. *Ocean and Coastal Management*, **30**(2-3), special issue: *Coastal Zone Management in the Mediterranean* (ed E. Özhan), 1996, 89-113.

Jeftic L., Integrated coastal zone management in the Mediterranean Action Plan of UNEP. In *Medcoast 93*, v.1, ed. E. Özhan, Medcoast Institute, Ankara, 1993, pp. 465-482.

Jeftic L., Keckes S. & Pernetta J.C., Implications on Future Climatic Changes for the Mediterranean Coastal Region. In *Climatic Change and the Mediterranean*, vol. 2, ed. L. Jeftic, S. Keckes & J.C. Pernetta, Arnold, London, 1996, pp. 1-26.

Jelgersma S. & Sestini G., Implications of a Future Rise in Sea-Level on the Coastal Lowlands of the Mediterranean. In *Climatic Change and the Mediterranean*, vol. 1, ed. L. Jeftic, L.D. Milliman & G. Sestini, Arnold, London, 1992, pp. 282-303.

Kliot N., Cooperation and Conflicts in Maritime Issues in the Mediterranean Basin. *GeoJournal*, **18**(3), 1989, 263-272.

Kliot N., Maritime boundaries in the Mediterranean: aspects of cooperation and dispute. In *Maritime Boundaries and Ocean Resources*, ed. G. Blake, Croom Helm, London, 1987, pp. 208-26.

Klomp R., Lessons learned from the experience with policy decisions and supporting tools in coastal water management? In *Medcoast 93*, v.1, ed. E. Özhan, Medcoast Institute, Ankara, 1993, pp. 35-52.

Langford S.R., Straight baselines and maritime boundary delimitation in the Mediterranean. In, *The Development of Integrated Sea-Use Management*, ed. H.D. Smith & A. Vallega, Routledge, London, 1991, 73-85.

Lanquar R., *Tourisme et environnement en Méditerranée. Enjeux et prospective*. Les Fascicules du Plan Bleu 8, Economica/Centre d'Action Régionale, Plan Bleu de la Méditerranée, Paris/Sophia Antipolis, 1995.

Le Houerou H.N., Vegetation and Land-Use in the Mediterranean Basin by the Year 2050: A Prospective Study. In *Climatic Change and the Mediterranean*, vol. 1, ed. L. Jeftic, L.D. Milliman & G. Sestini, Arnold, London, 1992, pp. 175-232.

Leanza U., Sico L., Ciciriello M.C., *Mediterranean continental shelf*. Oceana, New York, 1988.

Leanza U., The Regional System of Protection of the Mediterranean against Pollution. In *Il regime giuridico internazionale del Mare Mediterraneo*, ed. U. Leanza, Giuffré, Milan, Italy, 1987, pp. 381-412.

Lindh G., Hydrological and Water Resources Impact of Climate Change. In *Climatic Change and the Mediterranean*, vol. 1, ed. L. Jeftic, L.D. Milliman & G. Sestini, Arnold, London, 1992, pp. 58-93.

MAP/PAP, *Coastal area management programme. The coastal region of Syria*, Vol. I, *Synthesis report*, MAP Priority Actions Prohramme, Split, Croatia, 1992.

MAP/PAP, *Integrated management study for the area of Izmir*. MAP Priority Actions Programme, Split, Croatia, 1993.

Marchand H. *et al.*, *Les forêts méditerranéennes. Enjeux et perspectives*. Economica, Paris, 1990.

Margat J., L'eau dans le bassin méditerranéen. Situation et prospective. Economica, Paris, 1992.

Mariño M.G., Implications of Climatic Change on the Ebro Delta. In *Climatic Change and the Mediterranean*, vol. 1, ed. L. Jeftic, L.D. Milliman & G. Sestini, Arnold, London, 1992, pp. 304-327.

Mediterranean Action Plan, *Experience of Mediterranean historic towns in the integrated process of rehabilitation of urban and architectural heritage*. UNEP, Split, Croatia, 1987.

Mediterranean Action Plan, *State of the Mediterranean Marine Environment*. UNEP, Athens, 1989.

Micallef A., Socio-economic aspects of insular coastal management. In *Proceedings of the Second International Conference on the Mediterranean Coastal Environment Medcoast '95*, ed. E. Özhan, Middle East Technical University, Ankara, vol. 1, 1995, pp. 551-562.

Milliman J.D., Jeftic L. & Sestini G., The Mediterranean Sea and Climate Change. An Overview. In *Climatic Change and the Mediterranean*, vol. 1, ed. L. Jeftic, L.D. Milliman & G. Sestini, Arnold, London, 1992, pp. 1-14.

Milliman J.D., Sea-level Response to Climatic Change and Tectonics in the Mediterranean Sea. In *Climatic Change and the Mediterranean*, vol. 1, ed. L. Jeftic, L.D. Milliman & G. Sestini, Arnold, London, 1992, pp. 45-57.

Millot C., La circulation générale en Méditerranée occidentale. *Annales de Géographie*, **98**(549), 1989, 497-515.

Ministère de l'Environnement, *L'environnement méditerranéen. Contribution française*, Ministère de l'environnement, Paris, 1995.

Ministère de l'Environnement, *La protection de l'environnement méditerranéen*. Ministère de l'environnement, Paris, 1994.

Nicholls R.J. & Hoozemans F.M.J., Vulnerability to sea-level rise with reference to the Mediterranean region. In *Proceedings of the Second International Conference on the Mediterranean Coastal Environment MEDCOAST'95*, ed. E. Özhan, Middle East Technical University, Ankara, vol. 2, 1995, pp. 1199-1213.

Nir Y. & Elimelech A., Twenty five years of development along the Israeli Mediterranean coast: goals and achievements. In *Recreational Uses of Coastal Areas*, ed. P. Fabbri, Kluwer Academic Publishers, Dordrecht, Germany, 1990, pp. 211-218.

Nurit K., Cooperation and Conflicts in Maritime Issues in the Mediterranean Basin. *GeoJournal*, 18(3), 1989, 263-72.

O'Reilly G., Gibraltar et le détroit. *Hérodote*, 57, 1990, 59-75.

Oliver P., *The fisheries resources of the Mediterranean*, Vol. I, *Western Mediterranean*. FAO, Rome, 1983.

Özhan E., Coastal zone management in Turkey. *Ocean and Coastal Management*, **30**(2-3), special issue: *Coastal Zone Management in the Mediterranean* (E. Özhan ed), 1996, 153-176.

Palutikof J.P. & Wigley T.M.L., Developing Climate Change Scenarios for the Mediterranean Region. In *Climatic Change and the Mediterranean*, vol. 2, ed. L. Jeftic, S. Keckes & J.C. Pernetta, Arnold, London, 1996, pp. 27-56.

Pavasovic A., Coastal Zone Management: Experience of Mediterranean Action Plan-UNEP. In *Semienclosed Seas. Exchange of Environmental Experiences between Mediterranean and Caribbean Countries*, ed. P. Fabbri & G. Fierro, Elsevier Applied Science, London, 1992, pp. 110-150.

Pavasovic A., Integrated Coastal Management in the Mediterranean: Present State, Problems and Future. In *Regional seas towards sustainable development*, ed. S. Belfiore, M.G. Lucia & E. Pesaro, Franco Angeli, Milan, Italy, 1996, pp. 119-133.

Pavasovic A., The Mediterranean Action Plan phase II and the revised Barcelona Convention: new prospective for integrated coastal management in the Mediterranean region. *Coastal and Ocean Management*, special issue *Sustainable development ot the regional level: the Mediterranea* (ed. A. Vallega), **31**(2-3), 1996, 133-182.

Pavicevic B.S., Priority action 'Land-use planning in earthquake zones'? In *Land use planning in earthquake zones. Seismic risk reduction in the Mediterranean region. Documents produced in the first stage of the Priority Action (1984-1986)*, MAP Technical Reports Series No. 17, UNEP, Priority Actions Programme, Regional Activity Centre, Split, Croatia, 1987, pp. 1-12.

Perissoratis C., Georgas D. *et al.*, Implications of Expected Climatic Changes for the Island of Rhodes. In *Climatic Change and the Mediterranean*, vol. 2, ed. L. Jeftic, S. Keckes & J.C. Pernetta, Arnold, London, 1996, pp. 59-142.

Pirazzoli P.A., Sea-level changes in the Mediterranean. In *Sea-level changes*, ed. M.J. Tooley & I. Shennan, Blackwell, Oxford, UK, 1987, pp. 152-181.

Plan d'Action pour la Méditerranée, Ville de Marseille, *Protection du patrimoine archéologique sous-marin en Méditerranée*. A.M.P.H.I., Marseille, France, 1995.

Raftopoulos E., Sustainable Development and the Mediterranean Action Plan Regime: Legal and Institutional Aspects. In *Regional seas towards sustainable development*, ed. S. Belfiore, M.G. Lucia & E. Pesaro, Franco Angeli, Milan, Italy, 1996, pp. 197-229.

Raftopoulos E., Sustainable Development and the Mediterranean Action Plan Regime: Legal and Institutional Aspects. In *Regional seas towards sustainable development*, ed. S. Belfiore S., M.G. Lucia & E. Pesaro, Franco Angeli, Milan, Italy, 1996, pp. 197-229.

Ramade *et al.*, *Conservation des écosystèmes méditerranéens. Enjeux et perspectives*. Economica, Paris, 1990.

Ravagnan G., Productive development of lagoonal zones: available technologies and operational strategy. In *Management of living resources in the Mediterranean coastal area*, FAO, Rome, 1981, pp. 173-239.

Rehault J.-P., Boillot G. & Mauffret A., The Western Mediterranean Basin. In *Geological Evolution of the Mediterranean Basin*, ed. D.J. Stanley & F.-C. Wezel, Springer-Verlag, New York, 1985, pp. 101-130.

Republic of Tunisia, Ministry of Environment and Land Use Planning, *The Tunis Conference on Sustainable Development in Mediterranean. Agenda Med 21*. MAP/UNEP, Tunis, 1994.

Republic of Tunisia, Ministry of Environment and Land Use Planning, *The Tunis Conference on Sustainable Development in Mediterranean. The Conference Report*. MAP/UNEP, Tunis, 1994.

Ridolfi G., Il mare proibito. Profilo geografico degli usi militari del Mediterraneo. *Rivista Geografica Italiana*, **95**(2), 1988, 121-150.

Roucounas E., Sub-Marine Archaeological Research: Some Legal Aspects. In *Il regime giuridico internazionale del Mare Mediterraneo*, ed. U. Leanza, Giuffré, Milan, Italy, 1987, pp. 309-336.

Ruiz J.J., International Law Facing Mediterranean Sustainable Development: The Revision of the Barcelona Convention and Its Related Protocols. In *Regional seas towards sustainable development*, ed. S. Belfiore S., M.G. Lucia & E. Pesaro, Franco Angeli, Milan, Italy, 1996, pp. 230-255.

Sammak A.A., Coastal zone management in Egypt: present status and response options. In *Proceedings of the Second International Conference on the Mediterranean Coastal Environment Medcoast'95*, ed. E. Özhan, Middle East Technical University, Ankara, vol. 1, 1995, 465-477.

Sánchez A. *et al.*, Impacts of sea-level rise on the Ebro Delta: a first approach. *Ocean and Coastal Management*, **30**(2-3), 1996, 197-216.

Sanchez-Arcilla A. *et al.*, Impacts of sea-level rise on the Ebro Delta: An approach. In *Medcoast 93*, vol. 2, ed. E. Özhan, Medcoast Institute, Ankara, 1993, pp. 1103-1117.

Schmidt U.W., Ecological, Social and Financial Aspects of Coastal Aquaculture in the Mediterranean Region. In *Environmental aspects of Aquaculture Development in the Mediterranean Region*, UNEP/MED/Action Plan, Split, Croatia, 1987, pp. 33-40.

Scovazzi T., International Law of the Sea as Applied to the Mediterranean. *Ocean and Coastal Management*, **24**(1), 1994, 71-84.

Scovazzi T., Management regimes and responsibility for international straits: with special reference to the Mediterranean Straits. *Marine Policy*, **19**(2), 1995, 137-152.

Sestini G., Global warming, climatic changes and the Mediterranean. In *Symposium Mediterranean Seas 2000*, ed. N. Della Croce, Istituto di Scienze Ambientali Marine, Università di Genova, S. Margherita Ligure, Italy, 1993, pp. 373-393.

Sestini G., Implications of Climatic Changes for the Nile Delta. In *Climatic Change and the Mediterranean*, vol. 1, ed. L. Jeftic, L.D. Milliman & G. Sestini, Arnold, London, 1992, pp. 535-601. *Quoted as 1992/a.*

Sestini G., Implications of Climatic Changes for the Po Delta and Venice Lagoon. In *Climatic Change and the Mediterranean*, vol. 1, ed. L. Jeftic, L.D. Milliman & G. Sestini, Arnold, London, 1992, pp. 428-494. *Quoted as 1992/b.*

Sestini G., Jeftic L. & Milliman J. D., *Implications of expected climate changes in the Mediterranean region: an overview.* MAP Technical Reports Series. No 27, UNEP, Athens, 1989.

Sestini G., Sea Level Rise in the Mediterranean Region: Likely Consequences and Response Options. In *Semi-enclosed Seas. Exchange of Environmental Experiences between Mediterranean and Caribbean Countries*, ed. P. Fabbri & G. Fierro, Elsevier Applied Science, London, 1992, pp. 79-109.

Suárez De Vivero J.L., Ocean Issues and Management Tools in Spain: The Southwest Atlantic Waters. In *The Ocean Change: Management Patterns and the Environment*, ed. J. L. Suarez de Vivero, IGU Commission on Marine Geography and Universidad de Sevilla, Sevilla, Spain, 1992, pp. 119-128.

Tabet-Aoul E.H., Mediterranean coastal zone management efforts and aspects. In *Proceedings of the Second International Conference on the Mediterranean Coastal Environment Medcoast'95*, ed. E. Özhan, Middle East Technical University, Ankara, vol. 1, 1995, pp. 523-532.

The Mediterranean Institute, *The Mediterranean in the New Law of the Sea.* Foundation for International Studies, Malta, 1987.

Treves T., International Legal Problems of Marine Scientific Research with Particular Reference to the Mediterranean Sea. In *Il regime giuridico internazionale del Mare Mediterraneo*, ed. U. Leanza, Giuffré, Milan, Italy, 1987, pp. 289-308.

Tyrakowski K., The Role of Tourism in Land Utilization Conflicts on the Spanish Mediterranean Coast. *GeoJournal*, **13**(1), 1986, 19-26.

Udias A., Seismicity of the Mediterranean Basin. In *Geological Evolution of the Mediterranean Basin*, ed. D.J. Stanley & F.C. Wezel, Springer-Verlag, New York, 1985, pp. 55-64.

Ulrich W. S., Ecological, social and financial aspects of coastal aquaculture in the Mediterranean region. In *Environmental aspects of aquaculture development in the Mediterranean region. Documents produced in the period 1985-1897*, MAP Technical Reports Series No. 15, UNEP, Priority Actions Programme, Regional Activity Centre, Split, Croatia, 1987, 33-40.

UNEP, *State of the Mediterranean Marine Environment.* MAP Technical Reports Series No. 28, UNEP, Athens, 1989.

UNEP, *Arab cooperation for the protection and development of the marine environment and coastal areas resources of the Mediterranean.* UNEP Regional Seas Reports and Studies No. 52, UNEP, Athens, 1984.

UNEP, *Assessment of the State of Pollution of the Mediterranean Sea by Petroleum Hydrocarbons.* UNEP, Athens, 1988.

UNEP, *Avenirs du Bassin Mediterraneen. Environnement - Developpement 2000 - 2025.* UNEP, Athens, 1988.

UNEP, *Bibliography on effects of climate change and related topics.* MAP Technical Reports Series No. 29, UNEP, Athens, 1989.

UNEP, *Co-ordinated Mediterranean Pollution Monitoring and Research Programme (Med Pol - Phase I). Final Report 1975-1980.* UNEP, Athens, 1986.

UNEP, *Coastal Water Quality Control (Med. Pol. VII). Final reports of principal investigators.* UNEP, Athens, 1986.

UNEP, *Convention for the Protection of the Mediterranean Sea against Pollution and its related protocols.* United Nations, New York, 1982.

UNEP, FAO, *Assessment of the State of Pollution of the Mediterranean Sea by Mercury and Mercury Compounds.* UNEP, Athens, 1987.

UNEP, FAO, *Research on the Effects of Pollutants on Marine Organisms and their Populations (Med Pol IV). Final Reports of Principal Investigators.* UNEP, Athens, 1986.

UNEP, FAO, WHO, IAEA, *Assessment of the state of pollution of the Mediterranean Sea by organophosphorus compounds.* MAP Technical Reports, No. 58, UNEP, Athens, 1991.

UNEP, *Guidelines for Integrated Management of Coastal and Marine Areas - with Special Reference to the Mediterranean Basin*. UNEP Regional Seas Reports and Studies No. 161. PAP/RAC (MAP-UNEP), Split, Croatia, 1995.

UNEP, *Integrated management study for the area of Izmir*. MAP Technical Reports Series, 84, UNEP Regional Activity Centre for Priority Actions Programme, Split, Croatia, 1994.

UNEP, IOC, *Assessment of the state of pollution of the Mediterranean Sea by Petroleum Hydrocarbons*. MAP Technical Reports Series No. 19, UNEP, Athens, 1988.

UNEP, IUCN, *Directory of Marine and Coastal Protected Areas in the Mediterranean Region*. MAP Technical Reports Series No. 26, UNEP, Athens, 1989.

UNEP, Land-based pollution protocol to Cartagena Convention. *Marine Policy*, **16**(1), 1992, 43-49.

UNEP, *Long-term programme for pollution monitoring and research in the Mediterranean (MED POL), Phase II*. UNEP Regional Seas Reports and Studies No. 28, UNEP, Athens, 1983.

UNEP, *Mediterranean Action Plan and the Final Act of the Conference of Plenipotentiaries of the Coastal States of the Mediterranean Region for the Protection of the Mediterranean Sea (Barcelona, 2-16 February, 1976)*. UNEP, Nairobi, 1976.

UNEP, *Preliminary assessment of the state of pollution of the Mediterranean Sea by zinc, copper and their compounds and proposed measures*. UNEP, Athens, 1993.

UNEP, *Promotion of soil protection as an essential component of environmental protection in Mediterranean coastal zones. Documents produced in 1985-87*. MAP Technical Reports Series No.16, UNEP PAP/RAC, Split, Croatia, 1987.

UNEP, *Protocol concerning Mediterranean specially protected areas*. UNEP, Athens, 1986.

UNEP, *State of the Marine and Coastal Environment in the Mediterranean Region*. UNEP(OCA)/MED IG.5/Inf.3, UNEP, Athens, 1995 (draft).

UNEP (1995) *Guidelines for Integrated Management of Coastal and Marine Areas - with Special Reference to the Mediterranean Basin*. UNEP Regional Seas Reports and Studies, No. 161. Split, PAP/RAC (MAP-UNEP), 1995.

UNEP/MAP, *Action Plan for the Protection of the Marine Environment and the Sustainable Development of the Coastal Areas of the Mediterranean (MAP Phase II)*. Document UNEP(OCA)/MED IG.5/8, UNEP, Athens, 1995.

UNEP/MAP, *Final Act of the Conference of Plenipotentiaries on the amendments to the Convention for the Protection of the Mediterranean Sea against Pollution, to the Protocol for the Prevention of Pollution of the Mediterranean Sea by dumping from Ships and Aircraft and on the Protocol concerning Specially Protected Areas and Biological Diversity in the Mediterranean*. Document UNEP(OCA)/MED IG.6/7, Athens, UNEP, 1995.

UNEP/MAP, *Report of the Ninth Ordinary Meeting of the Contracting Parties to the Convention for the Protection of the Mediterranean Sea against Pollution and its Protocols*. Document UNEP(OCA)/MED IG.5.16, UNEP, Athens, 1995.

UNEP/WHO, *Biogeochemical Cycles of Specific Pollutants (Activity K). Final Report on project on survival of pathogenic microorganisms in seawater*. MAP Technical Reports Series, No. 55, UNEP, Athens, 1991.

University of Split, *Project "Environmental management of the Kastela Bay". Synthesis of the first research cycle 1988-1993*. CIMIS Centre, Split, Croatia, 1994.

Vallega A., Integrated Coastal Area Management in the Framework of the Barcelona Convention: Tool or Goal?. In *Regional seas towards sustainable development*, ed. S. Belfiore S., M.G. Lucia & E. Pesaro, Franco Angeli, Milan, Italy, 1996, pp. 168-177.

Vallega A., Regional level implementation of Chapter 17: the UNEP approach to the Mediterranean. *Ocean and Coastal Management*, special issue: *Earth summit implementation: progress achieved on oceans and coasts* (ed B. Cicin-Sain & R.W. Knecht), **29**(1-3), 1995, 251-278.

Vallega A., The Management of the Mediterranean Sea: The Role of Regional Complexity. *Ocean and Coastal Management*, **18**(2-4), 1992, 279-290.

Vallega A., Towards the sustainable management of the Mediterranean Sea. *Marine Policy*, **19**(1), 1995, 47-64.

Vallega, A., The Mediterranean after the 1995 Convention. The historical sense of a turnaround point. In *Proceedings of the Second International Conference on the Mediterranean Coastal Environment, Medcoast '95*, ed. E. Özhan. Medcoast Secretariat, Middle East Technical University, Ankara, Turkey, pp. 719-32.

Van Der Meulen F. & Salman A.H.P.M., Management of Mediterranean coastal dunes. Ocean and Coastal Management, **30**(2-3), 1996, 177-195.

Vratusa A., The Impact of Changes in Europe on the Mediterranean. In *Ocean Yearbook 10*, ed. Mann Borgese, N. Ginsburg & J.R. Morgan, University of Chicago Press, Chicago, Illinois, USA, 1993, p. 214-228.

Vukas B., The Protection of the Mediterranean Sea against Pollution. In *Il regime giuridico internazionale del Mare Mediterraneo*, ed. U. Leanza, Giuffrré, Milan, Italy, 1987, pp. 413-438.

Wigley T.M.L., Future Climate of the Mediterranean Basin With Particular Emphasis on Changes in Precipitation. In *Climatic Change and the Mediterranean*, vol. 1, ed. L. Jeftic, L.D. Milliman & G. Sestini, Arnold, London, 1992, pp. 15-44.

INDEX

Geographical Index

Author Index

258 INDEX

Subject Index

GENERAL

Administrative areas
administrative domain 75-7
property regime 76-7

Agenda 21
general features 5-7
Chapter 17 5-7
integrated coastal management 14-7

Archipelagic waters
baselines 82
legal regime 93-4

Baselines
normal baselines 78-80
straight baselines 78-80
seaport installations 81-2
bays 80
See also: bays

Bays
juridical 80
historical 80-1

Coastal area
concept 17-9
See also: coastal zone

Coastal area delimitation
conventional design 124-8
physical criteria 125-6
economic criteria 126
administrative criteria 126-7
jurisdictional criteria 126-7
ICM-consistent delimitation 128-30
ecosystem's relevance 128-30
sea land interface lines 130-2
depth lines 132-4
distance lines 134-5
rational landward delimitation 135-7
rational seaward delimitation 135-9

Coastal ecosystem
biotic community 60-3
abiotic niche 53-60
protection 62-3

within the coastal system 63-4
management 71-2
boundaries 72-4

Coastal management
concept 8-12
evolution 8-12
role of the UN 8-12
factors 8-12
See also: integrated coastal management

Coastal management objectives
final objectives 184-6
Agenda 21 184-6
contingent objectives 184-6
decision-making system 189-91
approach by planners and managers 202

Coastal management programmes
technical language 203-4
process 204-31
triggering 204-8
justification 208-12
initiation 213
preparation 214-22
planning 222-4
implementation 225-7
monitoring 228-31
evaluation 228-31
political relevance 232

Coastal morphology
types of coasts 55-7

Coastal organisation
concept 51-3

Coastal organisation
concept 99-100
coastal organisation model 113-6
See also: Coastal Organisation Model

Coastal organisation core sectors
concept 100; 123-4
urban pressure 104-6
tourism 106-7
industry 107-9
nodes 109-10
gateways 109-10
biomass use 110-12

integration 191-6
matrix-based assessment 196-200
decision-making processes 201-2

Exclusive Economic Zone
legal regime 91-2
coastal management 91-2
fisheries 91-3

Exclusive Fishery Zone
legal regime 92-3

Global change
concept 31-4
climate change 31-4
social change 32-3

Human Dimensions of Environmental Global
Change Programme
relevance to global change 32-3
goals 32-3

Human Dimensions Programme
See: Human Dimensions of Environmental
Global Change Programme

Integrated coastal management
concept 14-7
leading principles 14-6
approaches from literature 17
approaches from intergovernmental
organisations 17
See also: coastal management

International Geosphere-Biosphere Programme
relevance to global change 32-3
goals 32-3

Jurisdictional zones
jurisdictional domain 75-8
management regime 75-8; 96-7
See also: baselines, territorial sea

Large Marine Ecosystem
concept 64-66
relevance to management 64-66

Legal frameworks
relevance 187-8
hierarchical structure 187-8

Marine column
physical features 59-60
upwelling 60

Sustainable coastal development
concept 8-12
ecosystem integrity 12
economic efficiency 13
ethical values 13

Tectonic plates
dynamics 53-5
types of plates 53-5
life cycle of the oceans 55
types of margins 55-7
continental shelf 56-7
continental slope 57
continental rise 57
See also: continental shelf, coastal morpholog

Territorial sea
legal regime 82-5
coastal management 82-5

UN Commission on Sustainable Development
establishment 191-6
role 191-6

Use-ecosystem matrix
concept 156-8
methodology 156-8
See also: use-use matrix

Use-use matrix
concept 152-6
Global Marine Interaction Model 152-6
methodology 152-6
See also: use-ecosystem matrix

HE MEDITERRANEAN

astal area management
   coastal area delimitation 139-42
   programmes 27-30
   role of PAP/RAC 27-30

astal change
   general features 42-3
   impacts on soils 44-6
   water management 44-6

astal ecosystem
   concept 65-7
   extent 66-7
   local ecosystems 68-70

astal organisation
   core sectors 99-104
   urban pressure 104-6
   tourism 106-7
   industry 107-9
   nodes 109-10
   gateways 109-10
   biomass use 110-12
   fisheries 110-12
   aquaculture 110-12
   cultural heritage management 112-13
   coastal organisation model 113-7
   utility for management 113-7

astal organisation model
   construction methodology 113-7
   utility for management 113-7

obal change
   main data 21-2
   biological diversity 21-2
   economic and social diversity 22-3
   cultural diversity 23
   stage-based view 34-6
   trans-industrial stage 36-40
   coastal management 39-40
   climate change 40-2
   basic questions 46-9

risdictional zones
   territorial seas 94-6
   contiguous zones 94-6
   continental shelves 87-91; 94-6

   exclusive economic zones 91-3
   exclusive fishery zones 94-6

Legal framework
   features 187-8
   decision-making systems 191-6

MAP
   See: Mediterranean Action Plan

Mediterranean Action Plan
   history 24-7
   Barcelona convention 24-7
   organisation 24-7

Mediterranean Commission on Sustainable
   Development
   establishment 191-6
   role 191-6

Mediterranean Large Marine Ecosystem
   concept 65-6
   extent 65-6

NOTES

[1] The World Commission on Environment and Development, chaired by Gro Harlem Brundtland, was established as a consequence of the UN General Assembly resolution No. 38/161, adopted at the 38th Session of UN in the fall of 1983. The Commission functioned as an independent body. The Commission's report was discussed by UNEP's Governing Council and then was submitted to the General Assembly of UN during the 42nd Session in the fall of 1987. The report was entitled *Our Common Future* (Oxford, Oxford University Press, 1987).

[2] The Preparatory Committee (PreCom) of UNCED met four times (PreCom I, 1990, New York and Nairobi; PreCom II and III, March and August 1991, Geneva; PreCom IV, March-April 1992, New York). The discussion materials on Chapter 17 can be found in document A/Conf.151/PC/100/Add.21. The whole proceedings of the Conference, including those of the Preparatory Committee and complementing materials, were published in *Agenda 21 & the UNCED proceedings*, N.A. Robinson, P. Hassan and F. Burhenne-Guilmin eds, New York, Oceana Publications, 1992—1993.

[3] Teleology (from Greek *telos*, end, and *logia,* speech, reasoning) is that part of philosophy concerned with the study of ends.

[4] Moving from the framework of the *Man and Biosphere Programme* (MAB) of UNESCO, Young M.D. (1992) provided a general approach to sustainable development, focusing on economic efficiency. That approach could be usefully considered as a basis for dealing with sustainable coastal development.

[5] Wide discussion on the meaning and objective of ICM is provided by the special issue Integrated Coastal Management (Cicin-Sain B., ed) of *Ocean and Coastal Management*, 1993. In particular, the analysis of the conceptual and political links between sustainable development and integrated management is discussed in depth.

[6] This concept will be re-considered in Chapter 6 focusing on the coastal area organisation.

[7] This approach is rooted to the theory of complexity and, in particular, to the contributions by Le Moigne (1997-1994; 1990, especially Chapter 5). The theoretical approach to the interaction between human communities and the ecosystem is sensitive to the theory of Morin (1977, 1986). The transfer of this epistemology to the regional scale of ecosystem's management is discussed by Vallega A. (1995, especially Chapter 3).

[8] A detailed view of the history and present organisation of MAP, with special reference to the approach to coastal management, is provided by Pavasovic A. (1996).

[9] The greenhouse effect is due to increasing atmospheric concentration of carbon dioxide (CO_2), methane (CH_4), nitrous oxide (N_2O), ozone (O_3) and halocarbons (CFCs etc.).

[10] The IGBP programme consists of core projects. Coastal areas are specifically concerned with the core project *Land-Ocean Interactions in the Coastal Zone* (LOICZ). Information on IGBP can be found in *Global Change Reports* (International Council of Scientific Unions, ICSU, Stockholm).

[11] Economic and social investigations on the Mediterranean human change in the last quarter of the twentieth century, and designing scenarios referred to 2025, have been carried out by the *Blue Plan*, a Regional Activity Centre of MAP. The results achieved in the late 1980s can be found in M. Grenon and M. Batisse, *Futures for the Mediterranean Basin. The Blue Plan*. Oxford University Press, London, 1989.

[12] The first Mediterranean EEC member states were France and Italy; this status was acquired by Greece in 1981, and by Spain and Portugal in 1986.

[13] The Blue Plan is a Regional Activity Centre (BP/RAC) of MAP appointed to carry out environmental, economic and social investigations on the Mediterranean with the principal aim of constructing useful scenarios to operate the Barcelona Convention.

[14] In the framework of MAP, and with close reference to the IGBP approach, a wide set of contributions was provided on the consequences of climate change in the Mediterranean. They are included in two volumes: *Climatic Change and the Mediterranean*, vol. 1 (L. Jeftic, L.D. Milliman and G. Sestini G. eds), Arnold, London, 1992; vol. 2 (L. Jeftic, S. Keckes and J.C. Pernetta eds), Arnold, London, 1996. At the

present state this work may be regarded as the most methodologically advanced approach to impacts from climate change on a regional sea with special consideration for the coastal areas.

[15] The results were presented in L. Jeftic, S. Keckes and J.C. Pernetta eds, *Climatic Change and the Mediterranean*, Vol. 2, Arnold, London, 1992, pp. 1–26.

[16] ODUM P. E., *Fundamentals of ecology*. Philadelphia, Saunders, 1971.

[17] What is happening along the East African Rift Valley is a significant example of the birth of a new ocean and, as a result, an example of the processes generating the continental margins. The edges of the tectonic trenches characterising the Rift Valley are the proto-continental margins.

[18] The role of decision-making systems and the methods by which they may be assessed will be presented and discussed in Chapter 9.

[19] The conventions on the law of the sea, such as the United Nations Convention of the Law of the Sea (UN LOS, 1982) use the term 'zone' to indicate the national maritime jurisdictional extents. Juridical literature use 'zone' and, less frequently, 'belt'.

[20] In some parts of the ocean salt waters are subject to a property regime. As an example, this occurs in some coastal waters of Australia, where indigenous peoples have property rights on marine areas to catch living resources. However, this regime, usually depending on ancient cultural patterns, is not diffused.

[21] The LOS Convention came into force in November 16, 1994.

[22] The conceptual difference between the baseline and coastline is clear when the straight baselines are considered. They radically differ from the coastline because they may even be traced some miles seaward from the coastline. On the contrary, the normal baselines tend to coincide with the coastline since they are traced in the land-sea interface.

[23] As a result of the war in the former Yugoslavian area (1991-1995), these coasts currently belong to Slovenia, Croatia, Bosnia and Herzegovina, and the Federation of Yugoslavia (Serbia and Montenegro).

[24] Historically, the concept of territorial waters originated in the controversy between maritime powers over the status of the sea in the formative period of modern international law in the seventeenth century. Although the doctrine that the sea, by its nature, must be free for all states was eventually upheld, most commentators did recognise that, as a practical matter, a coastal state needed to exercise some jurisdiction over the waters adjacent to its shores. Two different principles were presented: on the one hand, the principle according to which the area of jurisdiction should be limited to cannon-shot range; on the other hand, the principle according to which that area should consist of a much more extended belt of uniform width adjacent to the coast. In the late eighteenth century these concepts coalesced in a compromise view that proposed a fixed limit of 3 nautical miles (1 marine league, or 3.45 land miles, corresponding to 5.5 kilometres).

[25] The LOS Convention, Article 76.4, provided two options to trace the outer edge of the continental margin where the coastal or archipelagic state claims its jurisdictional continental shelf as encompassing the continental margin: 'The fixed points comprising the line of the outer limits of the continental shelf on the seabed (...) either shall not exceed 350 nautical miles from the baselines from which the breadth of the territorial sea is measured or shall not exceed 100 nautical miles from the 2,000 isobath (...)'.

[26] The LOS Convention was adopted in 1982 and was in force by 1994 (November 16). Its discipline for the continental shelf was widely applied by states before its entering into force.

[27] 'Where the same continental shelf is adjacent to the territories of two or more States whose coasts are opposite each other, the boundary of the continental shelf appertaining to such States shall be determined by agreement between them. In the absence of agreement, and unless another boundary line is justified by special circumstances, the boundary is the median line, every point of which is equidistant from the nearest points of the baselines from which the breadth of the territorial sea of each State is measured' (Article 6, 1).

[28] Because of the geopolitical changes that occurred in the Yugoslavian space, this agreement is now concerned with the independent states which replaced that federation.

[29] The EEZ was included by Agenda 21, Paragraph 17.3, in the ICM framework.

[30] The concepts of coastal system and coastal organisation were introduced in Chapter 1.6.

[31] The main data on this process were presented in Chapter 2.2. See also Appendix 5.1 at the end of this Chapter.

[32] The Mediterranean island endowment possesses organisational features and has undergone evolution which would require the design of special EOMCAs. This is not the case of large islands such as Sicily, Sardinia and Corsica in the western Mediterranean, Malta in the Central and Cyprus in the Eastern Mediterranean because to some extent their economic organisation can be brought back to those of the continental coastal areas. Focus should be put on small islands, where coastal organisation is based on fragile ecosystems and profound cultural roots. Up to the end of the ninetieth century their spatial organisation was based on traditional horticulture, farming and fisheries, and settlements consisting of small historical villages. During the first half of the twentieth century urban settlements expanded and multi-function seaports were built. Afterwards coastal urbanisation has grown, recreational harbours and small airports have diffused, many rural areas have disappeared and indigenous plant and animal species have been endangered. Urban patterns have had it better than traditional organisational patterns (Brigand 1991: 12).

[33] See Chapter 1.2 and Table 1.3

[34] It is useful to remember that the present evolution phase of coastal management is characterised by the definition of innovating goals on both the global and regional (multi-national) scales. Meanwhile, states are asked to define sets of goals on the national scale in keeping with those agreed through international and/or regional treaties, or other similar legal tools. As a final result, the more an individual state is involved in regional and international co-operation, the more the real self-determination power of local communities decreases. On the global scale the final end of coastal management was determined by UNCED, which, in accordance with Agenda 21, Chapter 17, decided that the sustainable development of coastal areas was to be pursued by adopting ICM. Since Chapter 17 was unanimously agreed upon by States participating in UNCED, sustainable development should therefore be placed at the top level of the *hierarchical spectrum of coastal management goals,* and regarded as the reference basis for setting up coastal use frameworks.

[35] This classification is only tendentially binary for two reasons. First, the category of "presence of relationships" is sub-divided into three groups (neutral, conflicting and beneficial) instead of only two. This is due to the need to also represent those cases where the two uses coexist without interacting (neutral relationship). Secondly, both the conflicting and beneficial relationships are sub-divided according to the *direction of interaction* so prefiguring three cases: (i) the inputs from the line to the column; (ii) the input from the column to the line; and (iii) associated inputs from the line to the column and *vice-versa.*

[36] These experiments took place in courses held between 1995 and 1997 in the Architecture Faculty of the University of Genoa, Italy.

[37] According to Miles (quoted by Cicin-Sain, 1992), who especially refers to the United States, the typical kinds of conflict between users have involved: (i) ocean uses *versus* environmental protection; (ii) fishing *versus* marine cables; (iii) nearshore-located activities; (iv) fishing *versus* environmental quality; (v) fishing *versus* military uses; (vi) offshore oil and gas development *versus* navigation. A breakdown of frequent conflicts, deduced by an extended international survey distinguishing developed and developing countries, is presented by Cicin-Sain and Knecht (1998, 232-235).

The GeoJournal Library

1. B. Currey and G. Hugo (eds.): *Famine as Geographical Phenomenon.* 1984
   ISBN 90-277-1762-1
2. S.H.U. Bowie, F.R.S. and I. Thornton (eds.): *Environmental Geochemistry and Health.*
   Report of the Royal Society's British National Committee for Problems of the Environment. 1985   ISBN 90-277-1879-2
3. L.A. Kosiński and K.M. Elahi (eds.): *Population Redistribution and Development in
   South Asia.* 1985   ISBN 90-277-1938-1
4. Y. Gradus (ed.): *Desert Development.* Man and Technology in Sparselands. 1985
   ISBN 90-277-2043-6
5. F.J. Calzonetti and B.D. Solomon (eds.): *Geographical Dimensions of Energy.* 1985
   ISBN 90-277-2061-4
6. J. Lundqvist, U. Lohm and M. Falkenmark (eds.): *Strategies for River Basin Management.* Environmental Integration of Land and Water in River Basin. 1985
   ISBN 90-277-2111-4
7. A. Rogers and F.J. Willekens (eds.): *Migration and Settlement.* A Multiregional Comparative Study. 1986   ISBN 90-277-2119-X
8. R. Laulajainen: *Spatial Strategies in Retailing.* 1987   ISBN 90-277-2595-0
9. T.H. Lee, H.R. Linden, D.A. Dreyfus and T. Vasko (eds.): *The Methane Age.* 1988
   ISBN 90-277-2745-7
10. H.J. Walker (ed.): *Artificial Structures and Shorelines.* 1988   ISBN 90-277-2746-5
11. A. Kellerman: *Time, Space, and Society.* Geographical Societal Perspectives. 1989
    ISBN 0-7923-0123-4
12. P. Fabbri (ed.): *Recreational Uses of Coastal Areas.* A Research Project of the Com-
    mission on the Coastal Environment, International Geographical Union. 1990
    ISBN 0-7923-0279-6
13. L.M. Brush, M.G. Wolman and Huang Bing-Wei (eds.): *Taming the Yellow River: Silt
    and Floods.* Proceedings of a Bilateral Seminar on Problems in the Lower Reaches
    of the Yellow River, China. 1989   ISBN 0-7923-0416-0
14. J. Stillwell and H.J. Scholten (eds.): *Contemporary Research in Population Geography.*
    A Comparison of the United Kingdom and the Netherlands. 1990
    ISBN 0-7923-0431-4
15. M.S. Kenzer (ed.): *Applied Geography.* Issues, Questions, and Concerns. 1989
    ISBN 0-7923-0438-1
16. D. Nir: *Region as a Socio-environmental System.* An Introduction to a Systemic
    Regional Geography. 1990   ISBN 0-7923-0516-7
17. H.J. Scholten and J.C.H. Stillwell (eds.): *Geographical Information Systems for Urban
    and Regional Planning.* 1990   ISBN 0-7923-0793-3
18. F.M. Brouwer, A.J. Thomas and M.J. Chadwick (eds.): *Land Use Changes in Europe.*
    Processes of Change, Environmental Transformations and Future Patterns. 1991
    ISBN 0-7923-1099-3

The GeoJournal Library

19. C.J. Campbell: *The Golden Century of Oil 1950–2050.* The Depletion of a Resource. 1991   ISBN 0-7923-1442-5
20. F.M. Dieleman and S. Musterd (eds.): *The Randstad: A Research and Policy Laboratory.* 1992   ISBN 0-7923-1649-5
21. V.I. Ilyichev and V.V. Anikiev (eds.): *Oceanic and Anthropogenic Controls of Life in the Pacific Ocean.* 1992   ISBN 0-7923-1854-4
22. A.K. Dutt and F.J. Costa (eds.): *Perspectives on Planning and Urban Development in Belgium.* 1992   ISBN 0-7923-1885-4
23. J. Portugali: *Implicate Relations.* Society and Space in the Israeli-Palestinian Conflict. 1993   ISBN 0-7923-1886-2
24. M.J.C. de Lepper, H.J. Scholten and R.M. Stern (eds.): *The Added Value of Geographical Information Systems in Public and Environmental Health.* 1995
                                                                     ISBN 0-7923-1887-0
25. J.P. Dorian, P.A. Minakir and V.T. Borisovich (eds.): *CIS Energy and Minerals Development.* Prospects, Problems and Opportunities for International Cooperation. 1993
                                                                     ISBN 0-7923-2323-8
26. P.P. Wong (ed.): *Tourism vs Environment: The Case for Coastal Areas.* 1993
                                                                     ISBN 0-7923-2404-8
27. G.B. Benko and U. Strohmayer (eds.): *Geography, History and Social Sciences.* 1995
                                                                     ISBN 0-7923-2543-5
28. A. Faludi and A. der Valk: *Rule and Order. Dutch Planning Doctrine in the Twentieth Century.* 1994   ISBN 0-7923-2619-9
29. B.C. Hewitson and R.G. Crane (eds.): *Neural Nets: Applications in Geography.* 1994
                                                                     ISBN 0-7923-2746-2
30. A.K. Dutt, F.J. Costa, S. Aggarwal and A.G. Noble (eds.): *The Asian City: Processes of Development, Characteristics and Planning.* 1994   ISBN 0-7923-3135-4
31. R. Laulajainen and H.A. Stafford: *Corporate Geography.* Business Location Principles and Cases. 1995   ISBN 0-7923-3326-8
32. J. Portugali (ed.): *The Construction of Cognitive Maps.* 1996   ISBN 0-7923-3949-5
33. E. Biagini: *Northern Ireland and Beyond.* Social and Geographical Issues. 1996
                                                                     ISBN 0-7923-4046-9
34. A.K. Dutt (ed.): *Southeast Asia: A Ten Nation Region.* 1996   ISBN 0-7923-4171-6
35. J. Settele, C. Margules, P. Poschlod and K. Henle (eds.): *Species Survival in Fragmented Landscapes.* 1996   ISBN 0-7923-4239-9
36. M. Yoshino, M. Domrs, A. DouguÇdroit, J. Paszynski and L.D. Nkemdirim (eds.): *Climates and Societies – A Climatological Perspective.* A Contribution on Global Change and Related Problems Prepared by the Commission on Climatology of the International Geographical Union. 1997   ISBN 0-7923-4324-7
37. D. Borri, A. Khakee and C. Lacirignola (eds.): *Evaluating Theory-Practice and Urban-Rural Interplay in Planning.* 1997   ISBN 0-7923-4326-3

The GeoJournal Library

38. J.A.A. Jones, C. Liu, M-K. Woo and H-T. Kung (eds.): *Regional Hydrological Response to Climate Change*. 1996   ISBN 0-7923-4329-8
39. R. Lloyd: *Spatial Cognition*. Geographic Environments. 1997   ISBN 0-7923-4375-1
40. I. Lyons Murphy: *The Danube: A River Basin in Transition*. 1997 ISBN 0-7923-4558-4
41. H.J. Bruins and H. Lithwick (eds.): *The Arid Frontier*. Interactive Management of Environment and Development. 1998   ISBN 0-7923-4227-5
42. G. Lipshitz: *Country on the Move: Migration to and within Israel, 1948–1995*. 1998
                                                       ISBN 0-7923-4850-8
43. S. Musterd, W. Ostendorf and M. Breebaart: *Multi-Ethnic Metropolis: Patterns and Policies*. 1998   ISBN 0-7923-4854-0
44. B.K. Maloney (ed.): *Human Activities and the Tropical Rainforest*. Past, Present and Possible Future. 1998   ISBN 0-7923-4858-3
45. H. van der Wusten (ed.): *The Urban University and its Identity*. Roots, Location, Roles. 1998   ISBN 0-7923-4870-2
46. J. Kalvoda and C.L. Rosenfeld (eds.): *Geomorphological Hazards in High Mountain Areas*. 1998   ISBN 0-7923-4961-X
47. N. Lichfield, A. Barbanente, D. Borri, A. Khakee and A. Prat (eds.): *Evaluation in Planning*. Facing the Challenge of Complexity. 1998   ISBN 0-7923-4870-2
48. A. Buttimer and L. Wallin (eds.): *Nature and Identity in Cross-Cultural Perspective*. 1999   ISBN 0-7923-5651-9
49. A. Vallega: *Fundamentals of Integrated Coastal Management*. 1999
                                                       ISBN 0-7923-5875-9
50. D. Rumley: *The Geopolitics of Australia's Regional Relations*. 1999
                                                       ISBN 0-7923-5916-X

KLUWER ACADEMIC PUBLISHERS – DORDRECHT / BOSTON / LONDON